Instructor's Manual
with Transparency Masters

TO ACCOMPANY

Technical Communication

EIGHTH EDITION

JOHN M. LANNON

University of Massachusetts, Dartmouth

LONGMAN

An Imprint of Addison Wesley Longman, Inc.

New York • Reading, Massachusetts • Menlo Park, California • Harlow, England
Don Mills, Ontario • Sydney • Mexico City • Madrid • Amsterdam

Sponsoring Development Manager: Arlene Bessenoff
Developmental Editor: David Munger, The Davidson Group
Supplements Editor: Donna Campion
Electronic Page Makeup: Dianne Hall, The Davidson Group

Instructor's Manual with Transparency Masters To Accompany *Technical Communication*, Eighth Edition by John M. Lannon.

ISBN: 0-321-06131-4

9798990001—CRS—987654321

Contents

Table of Master Sheets

PLEASE NOTE: *Selected Master Sheets from the Instructor's Manual are available via electronic file. The software is provided to adopters at no extra charge. For more information, please visit our Web site at http://www.awlonline .com/lannontech or write to:*

Addison Wesley Longman
The Higher Education Publishing Group
1185 Avenue of the Americas
New York, New York 10036

The Composition Teacher
as Technical Writing Teacher

As demand increases for technical writing courses, many instructors are recruited to teach a subject that they might regard as alien to their training, ability, and primary interests. But anyone experienced in teaching composition can make an easy and rewarding transition to teaching technical writing. Your proven ability to assess clarity, economy, organization, and rhetorical effectiveness provides the essential ingredient—along with a touch of curiosity and willingness to experiment. In this course, as in any composition course, purpose, audience, and rhetorical strategy are stressed.

In technical writing, a major rhetorical challenge is to write for an audience whose technical understanding is less than the writer's own. Accordingly, the emphasis in this text is on writing for a general audience. Instructors without technical background, therefore, make an ideal audience—as do students with widely varied majors.

In a technical writing class, you don't need to struggle for answers to the student's implied question on each assignment: "Why are we doing this?" Because students choose subjects with observable limits, and because they write for a specific reader in a specific situation, they are able to make the connection between writing in the classroom and writing in the workplace. And with high motivation, skills improve quickly.

Students learn to master rhetorical strategies by writing about subjects of primary or immediate interest. The issues are more substantive than abstract. A report analyzing why the campus has no day-care center may require these expository skills: classification, definition, description, narration, and persuasion, in addition to strategies for summary writing, outlining, primary and secondary research, and letter writing. Along with obtaining valuable writing practice, then, students in this course develop a clear sense of purpose, because they write about problems that touch them and their community. The range and variety of topics are infinite, with repeated emphasis on highly informative writing. Writing is taken out of the rarefied English classroom and based in the real world. As an act of communication for a specific purpose to a specific audience, writing becomes more a cognitive than an affective task, more than an exercise in creative self-expression. Justification for such assignments is both implicit and explicit. With practice in thinking and writing for a tangible situation and purpose, for an audience who will *use* the information, students in any major leave the course better prepared to think and write incisively about any subject.

A report-writing assignment is, in effect, an instructor's call to "teach me," rather than "discover yourself." The practical purpose for writing is always clear. Unlike the rhetorical errors in more personal writing, deficiencies in a factual message can be identified readily; moreover, a summary, an expanded definition, a set of instructions, a physical description, or a proposal provides common ground for student-teacher discussion of content, arrangement, and style.

For the skeptical newcomer, technical writing's greatest liability is its name. The term "technical," often misleading for both instructors and students, leads to misunderstanding about what goes on in a technical writing course. It is one thing to discuss a *technical subject* (a specialized subject, usually mechanical or scientific); it is another to discuss any subject, technical or not, from a *technical point of view* (an informed and precise perspective from which the writer sees the related particulars of a subject). Even the most abstract subjects are discussed from a technical point of view if interpretations and conclusions are predicated on demonstrable evidence, and if the writing has utility beyond self-expression; literary criticism is an example.

In technical writing, the cognitive tasks of observing, interpreting, and reporting discourage any tendency to make absolute or sweeping statements. And, because guidelines for structure and format include an explicit and inclusive title, a clear statement of purpose, a detailed outline, and relevant headings, students maintain a sense of direction consistent with purpose. Far from enforcing mindless, mechanical transcription, technical writing assignments elicit thought and expression that are deliberate; volition rather than chance shapes the message.

Because of its concrete subject matter, technical writing encourages analytical thought. Students learn to pose imaginative questions, to answer them by precisely interpreting factual evidence, and to communicate their findings in a "professional" format. The approach is empirical not mechanical. Students see that they are writing for a reason, and that good writing is the product of a good plan and a clear sense of the specific reader's specific needs. Written assignments, oral reports, and class discussions about analogues in the real world evaluating your college's remedial program, establishing a student-operated food co-op, comparing four popular wood-burning stoves, analyzing safety devices at a local nuclear power plant—all have practical translations, are easy to justify, and are carried out with enthusiasm. Ideally, a student report will also satisfy an assignment in another course.

As a major course project, the analytical report can evolve from shorter assignments in summary writing, definition, description, and the like. Students are motivated when convinced that they are not performing an exercise in busywork or philosophical rambling; instructors are pleased to learn something informative instead of suffering the usual, thankless and bleary-eyed plodding through unmemorable essays.

In short, teaching technical writing is one way in which instructors can make the required conceptual and practical adjustment from education for its own sake to education with a visible purpose. Such a change hardly means settling for second best. This kind of teaching, as many continue to discover, offers the occasion for growing professionally and for actively involving our students in reciprocal teaching and learning.

Using the Masters for
Dittos and Transparencies

This manual is designed in large format to accommodate varied masters from which copies of quizzes, writing samples, and syllabi can be made, or from which photocopies may be shown on an opaque projector.

For Quizzes

No book will do students any good unless they read it. To ensure that your students have (1) done the reading and (2) understood what they have read, you might use the quiz at the end of each chapter discussion section. Each quiz has ten objective questions that can be answered in five to ten minutes. To simplify reproduction, each full quiz occupies only one side of a page. You can reproduce the quizzes directly from this manual without retyping by using a Thermofax machine and ditto masters.

For Writing Samples

In addition to quizzes, many chapter discussions are supplemented by transparency (or Thermofax) masters of visuals and writing samples. In the discussions of the letter and short-report chapters, transparency masters of student writing illustrate successful responses to exercises in order to complement many of the on-the-job examples from the textbook. You can make transparencies (for use on an overhead projector) directly from the manual without retyping—use a 3M Transparency Maker or a similar machine. Or you can order (at no charge) the complete package of acetate transparencies directly from the publisher.

For Syllabi and Course Description

Either of the two sample syllabi, the course specifications, and the description of a grading system can be reproduced directly.

Advantages of a Visual Format

Besides enhancing class discussion and lectures and improving students' attention, routine exposure to opaque or overhead projection is valuable preparation for students' careers. Research suggests that, in any presentation, speakers who use visuals are regarded as better prepared than speakers without such aids.

How Master Sheets Are Distributed in This Manual

To follow the same principles of efficiency set forth in the textbook, I deliberately omit master sheets (except for quizzes) from some chapters. The bulk of masters is in Part I (to enhance discussions about the writing process) and in Part V (to provide guidance in planning and revising typical documents). For Part II, documents produced by your own students should provide abundant examples.

As a quick survey of the Table of Master Sheets suggests, the emphasis in this material is on the *process*, not just the *product.* Instead of merely showing sample responses to this or that assignment, many of the masters illustrate the writing process as a *thinking* process.

Annotated Bibliography
of Resources for Teachers

Journals

The Journal of Business Communication. Association for Business Communication, c/o Department of Speech, Baruch College, 17 Lexington Avenue, New York, NY 10010. Association membership includes the journal and *Business Communication Quarterly.* Student membership is available.

Technical Communication. Society for Technical Communication, 900 North Stuart St., Arlington, VA 22203. Membership includes the journal and the society's newsletter, *Intercom.* Student membership is available.

The Journal of Technical Writing and Communication. Baywood Publishing Company, 26 Austin Avenue, P.O. Box 337, Amityville, NY 11701.

Technical Communication Quarterly. Association of Teachers of Technical Writing, c/o Department of Rhetoric, University of Minnesota, St. Paul, MN 55108. This journal is an indispensable source of ideas, approaches, and information on the most current publications of interest to teachers. Student membership is available.

IEEE Transactions on Professional Communication. Institute of Electrical and Electronics Engineers. This journal carries useful articles on the theory and practice of technical communication.

Journal of Business and Technical Communication. Sage Publications, 2455 Teller Road, Thousand Oaks, CA 91320.

Communication Research. Sage Publications.

Journal of Employee Communication Management. Lawrence Ragan Communications, 212 W. Superior St., Chicago, IL 60610.

Bibliographies

An Annotated Bibliography on Technical Writing, Editing, Graphics, and Publishing 1966–1980, available from the Society for Technical Communication. This selected listing includes entries on education, word processing, and principles of communication.

Basic Texts in Technical and Scientific Writing, an annotated bibliography published by the Society for Technical Communication.

"Bibliography on Education in Technical Writing and Communication, 1978–1983," Myra Kogen and the CCCC Committee on Technical Communication. *Technical Communication* 31.4 (1984): 45–48.

"1997 ATTW Bibliography." *Technical Communication Quarterly* 8.1 (1999): 91–117. One of a series of yearly listings of literature in technical and scientific communication.

Research Sourcebooks

Document Design: A Review of the Relevant Research. Ed. Daniel B. Felker, et al. Washington, DC: American Institutes for Research, 1980. Recognizing that "the organization of a document may be just as important as its language," the editors examine research from various disciplines as it relates to design of documents.

Research in Technical Communication: A Bibliographic Sourcebook. Ed. Michael Moran and Debra Journet. Westport, CT: Greenwood, 1985. This comprehensive work summarizes research in the history, theory, practice, and teaching of technical communication.

Research Strategies in Technical Communication. Lynette R. Porter and William Coggin. New York: John Wiley, 1995. This is an essential reference for teachers and researchers.

Technical and Business Communication: Bibliographic Essays for Teachers and Corporate Trainers. Ed. Charles H. Sides. Urbana, IL: National Council of Teachers of English, 1989. This collection offers a broad, interdisciplinary view of technical communication.

Technical Communication Frontiers: Essays in Theory. Ed. Charles H. Sides. St. Paul: Association of Teachers of Technical Writing, 1994.

Writing in Nonacademic Settings. Ed. Dixie Goswami and Lee Odell. New York: Guilford, 1985. This collection covers workplace writing practices and rhetorical problem solving from varied perspectives.

Classroom Resources

The Case Method in Technical Communication. Theory and Models. Ed. John R. Brockman. St. Paul, MN: Association of Teachers of Technical Writing, 1985. This work is an excellent introduction to the use of real-world cases for teaching technical writing.

Courses, Components, and Exercises in Technical Communication. Ed. Dwight W. Stevenson. Urbana, IL: National Council of Teachers of English, 1981. This rich collection offers a broad array of teaching and motivational strategies.

Introduction to the Engineering Profession. M. David Burghardt. New York: Harper, 1991. Technical writing teachers with humanities backgrounds will find this overview engaging and invaluable.

The Teaching of Technical Writing. Ed. Donald H. Cunningham and Herman A. Estrin. Urbana, IL: National Council of Teachers of English, 1975. This early but perennially relevant anthology offers articles on theory, teaching strategies, and workplace needs.

Writing in Multicultural Settings. Eds. Carol Severino, Juan C. Guerra, and Johnella E. Butler. New York: MLA, 1997. Essential reading for any teacher in a culturally heterogeneous classroom.

General Resources

The Art of Thinking: A Guide to Critical and Creative Thought. 5th ed. Vincent R. Ruggiero. New York: Longman, 1998. Excellent and accessible coverage of creative and critical thinking.

Bugs in Writing: Debugging Your Prose. Lyn Dupré. Reading, MA: Addison-Wesley, 1995. An excellent reference for matters of grammar, page design, audience analysis, and persona, this book offers a no-nonsense approach.

Business Information: How to Find It, How to Use It. 3rd ed. Michael R. Lavin. Phoenix, AZ: Oryx, 1999. This outstanding work covers research design and a vast array of sources, and offers vital wisdom for evaluating and interpreting findings.

Designing Instructional Text. 2nd ed. James Hartley. London: Kogan Page, 1985. Structured as a manual, this work is an excellent source of advice for audience-centered page design and visuals.

Directions in Technical Writing and Communication. Jay R. Gould. Farmingdale, NY: Baywood, 1985. This collection of readings is designed to supplement textbooks and class discussion.

Editing: The Design of Rhetoric. Sam Dragga and Gwendolyn Gong. Amityville, NY: Baywood, 1989. This provocative and insightful book examines the editing process "not merely as a mechanical procedure, but a complex and creative process" with a basis in rhetorical theory. Includes principles and strategies for editing verbal and visual features.

Engineering Psychology and Human Performance. 2nd ed. Christopher D. Wickens. New York: Harper, 1992. A detailed analysis of "human factors" as they affect information processing, this work is essential reading for professional communicators.

Guidelines for Document Designers. Ed. Daniel B. Felker, et al. Washington, DC: American Institutes for Research, 1981. With its stated goal—making documents easier to read and understand—this invaluable work offers principles for structuring sentences, designing pages, and presenting numerical and qualitative information.

A History of Professional Writing Instruction in America: Years of Acceptance, Growth, and Doubt. Katherine H. Adams. Dallas, TX: Southern Methodist University Press, 1993. This broad review offers a rich context for teachers in our discipline.

Information Anxiety. Richard S. Wurman. New York: Doubleday, 1989. An architect/information designer offers an innovative perspective on coping with information overload, as well as advice on designing documents for appeal and access.

International Business Communication. David A. Victor. New York: HarperCollins, 1992. An excellent introduction to this topic.

New Essays in Technical and Scientific Communication. Ed. Paul V. Anderson, et al. Farmingdale, NY: Baywood, 1983. The twelve essays in this collection cover both theory and practice.

Professional Ethics Report. This fall quarterly newsletter is published by the American Association for the Advancement of Science, 1333 H Street, NW, Washington, DC 20005. Write and ask to be placed on the mailing list.

Technical Editing. Carolyn D. Rude. Needham Heights, MA: Allyn & Bacon, 1997. This excellent overview of editing as a reader-centered process covers copy editing, substantive editing, and management and production.

Technical Writing: Theory and Practice. Eds. Bertie E. Fearing and W. Keats Sparrow. New York: Modern Language Association, 1989. This broad selection of essays covers history, theory, corporate practices, and teaching strategies and resources.

Understanding Research Methods. 2nd ed. Gerald R. Adams and Jay D. Schvaneveldt. White Plains, NY: Longman, 1991. This work offers valuable insight into research design, methodology, and evaluation.

Visual Explanations: Images and Quantities, Evidence and Narrative. Edward R. Tufte. Chesire, CT: Graphics Press, 1997. An essential reference for matters of graphics and critical analysis of information.

General Suggestions

Background Reading

Because technical writing is (at least by one definition) applied rhetoric, a new instructor's preparation should build on a solid foundation in classical rhetoric. For this purpose, the most concise and comprehensive source I know of is Edward P. J. Corbett's *Classical Rhetoric for the Modern Student* (Oxford University Press).

Classroom Layout

A technical writing class works best in the workshop format. Try to have your course scheduled in a room with several tables large enough for students to work in small editing groups and have plenty of room for paper shuffling.

Scheduling

Although sometimes difficult to schedule, two meetings a week seem to work best for a workshop. Because technical writing students generally are well motivated, they will easily tolerate 75-minute classes. These longer periods provide more continuity to the small-group and full-class sessions.

Hardware

An opaque projector and a permanent screen are essential fixtures. These provide a convenient medium for class discussion of student papers and other specimens. Keep overhead projector on hand for transparencies.

One user of earlier editions suggests using a mimeograph machine if one is nearby. She carries a box of masters on which students can record group work, and then runs the sheets off during the break.

An ideal tool for demonstrating the writing process intact is a microcomputer with word-processing software and a large screen.

Guest Speakers

Invite speakers from business and industry (the director of communications at your local power company, or the head of a local engineering firm, for example). Companies that strive for good public relations, such as utilities or paper companies, are especially cooperative.

The Workshop Approach

Workshops focus on the texts that students themselves have produced. The workshop approach operates on the premise that students can evaluate someone else's writing better than their own. Designed to take students out of their traditionally passive roles, the workshop involves them actively in evaluating and discussing writing. It helps familiarize students with the challenge of writing for audiences other than their instructor.

When first drafts or revisions are due, ask students to proofread and edit each other's assignments, using the appropriate revision checklist at the end of each chapter as a guide. Ask for a detailed evaluation of each assignment, including specific suggestions for revision. To encourage use of the handbook in the Appendix and the style suggestions in Chapter 13, ask students to use the correction symbols (rear endsheet) for referring the writer to specific sections for mechanical and stylistic improvements. (You also might ask them to keep a journal of their most troublesome mechanical and stylistic errors and to submit the journal periodically with a brief progress report.)

If a general reading audience is assumed, groups at each table should be heterogeneous. If a more specialized audience is assumed, the groups should be as homogeneous as possible. Provide a situational context for each workshop:

- For heterogeneous groups: "Assume that you are a customer, executive, or client who needs this information for [the specific purpose for which the assignment is written]. Would the information in this report fully serve your needs? Is it well presented [format, style, mechanics, usage]? What is effective about this piece? What needs improvement ?"

- For homogeneous groups: "Assume you are a section head who has to approve this piece [instructions, product description, and so on], written by one of your staff, before it is published in a company manual or prospectus. What specific advice would you give the writer for revising and refining the document?"

After allowing enough time (20 to 25 minutes) for small-group editing, ask for one or two nominations for outstanding papers to be discussed by the entire class. Display these papers on the opaque projector and read them aloud. Invariably, other class members will have additional insights and suggestions for improvement. By discussing a paper already recognized as superior, you can avoid damaging the writer's ego. Emphasize the good before taking up the areas in need of improvement.

Finally, ask students to revise their papers at home, applying their editors' comments, *before* they submit them to you for grading. Have them turn in both their revisions and their edited drafts.

In addition to marginal notes, I require from editors a brief evaluation (one or two paragraphs) of the individual features of *content, arrangement, style,* and *page design.* All students initial their summaries and receive extra credit for consistently good editing.

NOTE: Expect some resistance to the workshop for the first few sessions. Initially, some students feel they have nothing useful to say about a piece of writing. But with cheerleading and guidance on your part, the whole business soon will run smoothly. In fact, once students become accustomed to this approach, you can save class time by asking them to edit classmates' papers at home.

Have students identify a specific audience and use for each assignment. To reinforce the workplace connection, begin early with samples of not-so-good writing from business and industry (memos, letters) that the class can edit together, using the opaque projector.

Here are more suggestions for helping the workshops run smoothly:

1. Give periodic quizzes to ensure that students have read and understood the assigned chapters. For a workshop to succeed, students need to know the assigned reading.

2. Ask students to specify (in writing) an audience and use for *each* document they submit.

3. Emphasize *repeatedly* that all editors should assume the role of the writer's stipulated audience.

4. You generally should not see first drafts. Ask students to submit their edited draft along with the final draft.

5. Because an uninformed audience usually is a writer's biggest challenge, heterogeneous editing groups generally are more effective than homogeneous groups.

6. For full-class discussion of edited documents, use only those nominated as *superior.*

7. Before having students revise at home, hold at least one full-class workshop on that type of document.

8. For variety, use transparencies from time to time.

9. The workshop's purpose is to *actively* involve students in evaluation and thinking. Don't hesitate to call on members of the silent majority for commentary during full-class sessions.

10. For motivation and perspective, frequently bring in samples of real-world documents, both good and bad—or, better yet, ask your students to submit samples they've collected.

Due Dates for Assignments

Students should be given specific due dates for first drafts (for workshop editing) and deadline dates for all revisions. It's a good idea to impose a limit of only one revision for the assignments you have corrected. Besides preserving your sanity, this arrangement helps you avoid the role of teacher-as-proofreader.

Folders

Ask each student to buy a rugged, briefcase-like cardboard folder for holding all assignments and revisions. This collected work comes in handy during individual conferences. It also ensures that material is retrievable for those assignments that are cumulative.

Conferences

Schedule frequent conferences. These meetings are especially important early in the semester for students selecting topics for analytical reports (or proposals), and are important late in the semester as they work on these reports.

Document Standards

Except for complex visuals (Chapter 14), require that all assignments be "camera-ready." Besides providing an occasion for editing and revising, typing helps students develop a sense of professionalism and anticipates formal requirements on the job.

Assignments prepared on a word processor should be printed on a letter-quality printer; dot-matrix print simply is too hard to read. Newer dot-matrix printers, however, have a "double-strike" or near-letter-quality mode that creates more readable copy.

Attendance Policy

A workshop arrangement requires regular attendance. Subtracting two points from the semester's total (see Grading Procedure, pages 16–18) for each unexcused absence beyond two or three helps keep everyone coming.

Using the Objective
Test Questions

On the final pages of this manual is a bank of objective test questions that supplement the chapter quizzes. Of course, improvement in students' writing is the true measure of their progress. But an objective test at midterm or at semester's end can be useful:

1. For instructors who choose not to give weekly quizzes, the test helps differentiate weaker writers who have given their *best* effort from those who have given minimal effort.

2. Early announcement of a test is likely to motivate some students to read the book carefully, instead of merely skimming the chapters and focusing on the models.

3. The test itself is an occasion for students to review—and, presumably, to absorb better—key material.

To accommodate the chapter sequences used by different instructors, all test questions are organized and labeled by chapter.

Grading Procedure

An informal-contract grading system (like the one outlined on Master Sheet 1) has several advantages:

1. People who write on the job are not graded C+ or B–. A workplace document is deemed unacceptable, acceptable, or superior.[1]

2. Technical students generally feel more comfortable with quantitative evaluations, that is, with the guidelines clearly spelled out. Instructors are hard-pressed to explain to students (and often to themselves) the subtle distinction between an A– and a B+. Students see the contract system as fairer, and with good reason.

3. These clear distinctions help simplify peer evaluation during editing sessions.

4. With a contract system, students can do as much or as little as they deem necessary to achieve the grade they desire.

5. By keeping track of their points, students know exactly where they stand at any stage in the course. This knowledge is very helpful during conferences and for planning revisions.

The following system has been used successfully and has received enthusiastic student and faculty endorsement.[2]

---

[1]For greater flexibility within this grading scheme, you might tell students they could receive a grade that falls between the numerical values listed (for example, three out of a possible four).

[2]My thanks to Richard Dozier, University of Idaho, who devised the original version of this system.

Master Sheet 1

Grading System and Course Specifications

On the basis of my evaluation, each assignment in this course will be classified in one of three categories:

SUPERIOR   A document that meets professional requirements: worthwhile content; sensible organization; readable style; and appropriate design, visuals, and mechanics.

ACCEPTABLE   A document that satisfies most of these requirements, or one that satisfies all these requirements, but contains a reasonable number of mechanical errors that can be corrected easily.

UNACCEPTABLE   A document that needs extensive revision to meet all the requirements, or that has the type or amount of mechanical, rhetorical, or design errors that would distract readers.

Point Values for Individual Assignments

Assignment	Unacceptable	Acceptable	Superior
1. Summary	U	1	2
2. Expanded Definition	U	2	4
3. Collaborative Project	U	1	2
4. Visuals	U	1	2
5. Proposal Memo	U	1	2
6. Inquiry Letter	U	1	2
7. Claim Letter	U	1	2
8. Adjustment Letter	U	1	2
9. Résumé and Application Letter	U	2	4
10. Mechanism Description	U	3	6
11. Instructions	U	3	6
12. Progress Report	U	1	2
13. Justification Report	U	2	4
14. Oral Summary	U	2	4
15. Formal Report (or Proposal)	U	8	16
POINT TOTALS:	**0**	**30**	**60**

Master Sheet 2

Point Grade Equivalents

Grade	Required Point Range
A	54–60
B	44–53
C	30–43
D	26–29
F	25 or below

The grade earned on the above scale counts for _____ percent of the final grade. Class participation, quality of editing, quizzes, and other projects count for the remainder.

Course Specifications

Success in this course calls for three essentials: (1) attending and participating *actively* in the class, (2) following directions, and (3) meeting deadlines.

Attendance:

Assignments and exercises are due for almost every class session. Many classes follow a workshop format, in which we edit and discuss the writing done by you and your colleagues. Regular attendance and active participation in class discussion are therefore mandatory. For each unexcused absence beyond three, two points will be subtracted from your semester total.

General Directions:

All assignments must be typed on rag-bond (nonerasable) paper. Work that cannot be read on the opaque projector will not be accepted. If you use a word processor, print your document on a letter-quality printer; dot-matrix print will not be accepted.

Please note that the mere act of revision does not, in itself, guarantee a higher grade. A grade will improve only when the revised version shows enough improvement to merit a higher evaluation.

For grading, drafts must be stapled to your revisions. Place your revision on top, and staple in the upper left corner. Keep all work in a folder, for review and conferences. You may revise five assignments (excluding the final report) after I've graded them. Unless otherwise instructed, submit each document with a detailed audience-and-use analysis (as shown on pages 28–31 of your text).

Deadlines:

Readings, exercises, and assignments must be completed by the dates in the syllabus. Drafts must be completed on the due date so that they can be edited and discussed in workshops. Revisions are due by the following class session. All rewrites must be turned in by_____. No late submissions will be accepted.

Because you have the whole semester to work on your final reports or proposals, I will not allow any course grade of Incomplete.

Sample Syllabi

Each syllabus offered here covers a rigorous—but realistic—schedule of activities and assignments, based on forty-five class meetings. The pages can be reproduced on Thermofax dittos without being retyped.

Syllabus A—Basic Approach

If your students have little technical background (as with career-education students, first-year students in any major in two- or four-year programs, and two-year technical students who will not often be expected to write long documents on the job) you might use some version of this syllabus. Because the textbook chapters are self-contained, you can easily modify the suggested sequence to suit your goals. The sequence of chapters is explained later, in the discussions in Parts I, II, III, IV, and V.

Students following this syllabus will work on the long report in teams. If you prefer not to assign a Chapter 8 Collaborative Project, exercises 2–20 in Chapter 8 offer practice in researching specialized sources.

Syllabus B—Accelerated Approach

If your students are juniors and seniors with substantial backgrounds, or sophomores in four-year programs that require many long reports, you might use a version of Syllabus B. The workload is heavy, but the results are gratifying.

Syllabus B differs from Syllabus A in that it yields these additional assignments: project proposal, progress report, justification report, and oral report.

Both syllabi have ungraded exercises for the opening sessions, to get students writing early without them worrying about being penalized for poor writing.

Library Tour

Whatever your approach, try to arrange a tour of your college library. Most students at any level need some hands-on introduction to the more specialized guides to literature, reference works, indexes, and abstracts.

Arrange for a demonstration of the OCLC electronic catalog, Infotrac™ (a disk-based retrieval service), BRS or Dialog (mainframe database retrieval services), Internet, and other electronic resources for research.

Master Sheet 3

Syllabus A

Weekly Assignments and Activities

1. *Introduction:* Discuss course goals, grading, workshop concept, team projects and final project, graphics and page-design requirements. Read Chapter 1; do exercise 1 and Collaborative Project. Read Ch. 2; do exercise 1; discuss samples shown on the opaque projector.

2. *The Information Problem:* Read Chapter 3; do exercise 1 and Collaborative Project; workshop. *The Persuasion Problem:* Read Chapter 4; do exercises 2 and 3.

3. *The Ethics Problem:* Read Chapter 5; do exercise 2. *The Collaboration Problem:* Read Chapter 6; do Collaborative Project 1; workshop. Read Chapter 7 in preparation for the research project. Begin work on exercise 1, Phase One. Read Chapter 25, pages 522–529. Look over the exercise.

4. *Style:* Read Chapter 13; do all exercises. LIST OF POSSIBLE TOPICS FOR RESEARCH PROJECT IS DUE. *Page Design:* Read Chapter 15. Take a library tour. Read assigned sections of Chapters 8 and 9.

5. *Summarizing Information:* Read Chapter 11; do exercises 1 and 5; workshop; revised summary and abstract due next class meeting. TOPIC AND TENTATIVE BIBLIOGRAPHY FOR RESEARCH PROJECT ARE DUE. *Definition:* Read Chapter 21; do exercises 1, 3, and 4 or Collaborative Project 2; workshop; revised definition due next class meeting.

6. *Organizing for Readers:* Read Chapter 12; do the exercises; help teams develop working outlines for the final project, using chalkboard and opaque projector. TENTATIVE OUTLINE FOR RESEARCH PROJECT IS DUE. SIGN UP FOR TEAM CONFERENCES ON RESEARCH PROJECT. *Page Design:* Review Chapter 15; do exercise 3.

7. *Visuals:* Read Chapter 14; do exercises 1, 8, 9, 11, 12, and 13; do Collaborative Project 3; workshops. Continue work on tentative outlines for final project. *Reviewing Findings:* Read Chapter 10, do exercise 2.

8. *Letters:* Read Chapter 19, pages 365–375; do exercise 7 in class. *Inquiry Letters:* Read pages 375–377; do exercise 2; workshop; revised inquiry due next class meeting; mail inquiry letters for research project. *Claim Letters:* Read pages 377–379; do exercise 3 or 8; workshop; revised claim letter due next class meeting. Begin work on Chapter 7, exercise 1, Phase 2.

Master Sheet 4

9. *Résumés:* Read pages 379–387 and 393–398; begin work on exercise 6 or Collaborative Project 1; first draft of résumé due next class; workshop; revised résumé due next class. INTERVIEW QUESTIONS AND QUESTIONNAIRE ARE DUE.

10. *Application Letters:* Read pages 387–393; compose the application and follow-up letters in response to exercise 6 or Collaborative Project 1; workshop; revision due next class session. DETAILED OUTLINE FOR RESEARCH PROJECT IS DUE; workshop on outlines. Review Chapter 7 and work on exercise 1, Phase 2. Read Chapter 25.

11. *Mechanism Description:* Read Chapter 22; do exercises 3 and 4 in class; group brainstorming workshop; do a description outline based on exercise 1 or on Collaborative Project 2; outline workshop; prepare the description based on a Collaborative Project; workshop; revised description due next class meeting.

12. *Instructions:* Read Chapter 23, pages 462–482; do exercises 1 and 2; do outline for instructions based on exercise 4; workshop; prepare the instructions; workshop; revised instructions due next class.

13. *Long Report:* Review Chapters 7, 8, and 9; begin work in Chapter 7 on exercise 1, Phase 3; workshops on material that is volunteered. *Supplements:* Read Chapter 16; discussion and workshop on supplements.

14. *Documentation:* Read Appendix A; discuss various documentation systems. *Research Project:* First draft of research report due; workshop. If you want me to read the best draft of your report, you must turn it in at the beginning of this week.

15. *Final Project:* Final draft of report due; proofreading workshop.

Master Sheet 5

Syllabus B

Weekly Assignments and Activities

1. *Introduction:* Discuss course goals, grading, workshop concept, team projects, graphics and page design requirements. Read Chapter 1; do exercise 1. Read Chapter 2; do exercise 1. *The Information Problem:* Read Chapter 3; do exercise 1.

2. *The Persuasion Problem:* Read Chapter 4; do exercise 3 or 4; workshop. *The Ethics Problem:* Read Chapter 5; do exercise 2 and Collaborative Project. Discuss final project (proposal or report). Read Chapter 24, pages 522–551 and Chapter 25, pages 522–529?. Look over exercise 4, Chapter 24 and the exercise in Chapter 25. *Collaborative Guidelines:* Read Chapter 6.

3. *Style:* Read Chapter 13; do all exercises. Read Chapter 7 in preparation for final project. LIST OF POSSIBLE TOPICS FOR FINAL PROJECT IS DUE.

4. *Summarizing Information:* Read Chapter 11; do exercises 1 and 5; workshop; revised summary and abstract due next class meeting. *Definition:* Read Chapter 17; do exercises 1, 3, and 4 or Collaborative Project 2; workshop; revised definition due next class meeting. TOPIC AND TENTATIVE BIBLIOGRAPHY FOR FINAL PROJECT ARE DUE.

5. *Organizing for Readers:* Read Chapter 12; do the exercises. TENTATIVE OUTLINE FOR FINAL PROJECT IS DUE. *Visuals:* Read Chapter 14; do exercises 1, 9, 11, 12, and 13; do Collaborative Project 3; workshop. SIGN UP FOR OFFICE CONFERENCES ON FINAL PROJECT.

6. *Page Design:* Read Chapter 15; do exercise 4; workshop on exercise 4. *Project Proposal:* Read Chapter 24, pages 497–499; do exercise 3; workshop; revised proposal for final project due next class meeting. *Letters:* Read Chapter 19, pages 365–375; do exercise 7 in class. *Reviewing Findings:* Read Chapter 10; do exercise 2.

7. *Inquiry Letters:* Read Chapter 19, pages 375–377; write letter based on exercise 2; workshop; revised inquiry letter due next class meeting; mail inquiry letters for final project. *Claim Letters:* Read Chapter 19, pages 377–379; do exercise 3 or 8; workshop.

Master Sheet 6

8. *Résumés:* Read Chapter 19, pages 379–387 and 393–398; begin scanning classified ads for a job you could fill once you graduate (you will submit the ad with your application letter); compose a résumé; workshop; revised résumé due next class meeting. *Application Letters:* Read Chapter 19, pages 387–393; compose the application and follow-up letters; workshop; revisions due next class session.

9. *Research:* Review Chapters 7, 8, and 9. *Progress Report on Final Project:* Read Chapter 18, pages 350–353; do exercise 6. DETAILED OUTLINE FOR FINAL PROJECT IS DUE. Do exercises 1, 2, and 3 in Chapter 9. Interview questions and questionnaire are due next class meeting. Workshop on final-project outlines.

10. *Mechanism Description:* Read Chapter 22; do exercises 3 and 4 in class; do a description outline based on exercise 1 or on Collaborative Project 2; outline workshop; prepare a description based on exercise 1 or on Collaborative Project 2; workshop; revised description due next class meeting.

11. *Instructions:* Read Chapter 23, pages 462–482; do exercises 1 and 2; do outline for instructions based on exercise 4 or on one of the Collaborative Projects; workshop; prepare the instructions; workshop; revised instructions due next class.

12. *Final Project:* Read Chapter 24 or 25; begin work toward a completed draft of the proposal or report; general workshops on outlines, report sections, and so on. *Justification Report:* Read Chapter 18, pages 349–350; do exercise 1, 2, 3, or 4; workshop; revised justification report due next class meeting.

13. *Documentation:* Read Appendix A, discuss various documentation systems. Sign up for oral summaries. *Supplements:* Read Chapter 16, discuss various supplements; workshops on material that is volunteered. If you want me to read your best draft of your proposal or long report, you must turn it in by the end of this week.

14. *Final Project:* Workshops on completed drafts of proposals and reports, including supplements. *Oral Summaries:* Read Chapter 26; each student presents a ten-minute summary with visuals.

15. *Oral Summaries and Loose Ends:* FINAL REVISION OF TERM PROJECT (WITH ALL SUPPLEMENTS) IS DUE.

1

Introduction to
Technical Communication

The main point in this chapter is that all professional writing is done for specific readers in specific situations, to communicate information that readers will use. The writer's primary purpose is not to express personal feelings or opinions—or simply to transmit factual information; instead, the writer's purpose is to shape that information for the particular uses of a specific audience. In this sense, the notion of "user-friendliness" applies not only to computer hardware, software, and documentation, but also to any document written for its readers' instrumental use.

To help students understand that this is not just another composition course, spend time discussing the differences between technical and nontechnical writing. You might bring in examples of technical writing, such as short pieces from *Scientific American* or operating instructions for an electric tool or appliance, and examples of nontechnical writing, such as expressive or mood pieces from *Reader's Digest* or other popular magazines, or newspaper feature articles purportedly objective but often dripping with sentimentality.

Because motivation and attitude are crucial in getting students to improve their writing (research shows that students write more effectively when the subject is engaging, and when their purpose for writing is clearly defined), you might wish to amplify the section on the value of technical skills with quotations from business, industrial, and technical magazines. A quick library search should yield ample material.

Ask students for a memo based on exercise 2 or 3, identifying the kinds of writing they will have to do or discussing their specific expectations for the course. Besides getting students to write immediately—giving you an idea of individual strengths and weaknesses—this assignment has two advantages: exercise 2 helps students confront writing as a task in the working world, rather than as a string of classroom exercises in busywork; exercise 3 helps them think about their specific needs (to the extent that they recognize them), makes them feel they have a voice in the focus and direction of the course, and may help you adjust course content more directly toward their needs and expectations.

The Team Project works well as an early exercise in eliciting, sorting, organizing, and presenting information for specified use by a specified audience—all typical workplace tasks for a technical communicator.

For an early introduction to memos as the common medium for written communication within organizations, you might distribute copies of Master Sheet 7.

A Good First- or Second-Day Exercise

To emphasize that technical writing calls for clear, precise, and richly descriptive language, you might use this simple exercise:

Hold up a dime and ask students to describe it (without benefit of visuals or analogies) in enough detail to present a clear picture to an uninformed reader. Ask the class to ignore the dime's function, as well as the engraving on the two flat surfaces, and to concentrate only on shape, dimensions, and materials. Limit the description to fifty words.

After some grumbling and head scratching, the class will produce such specimens as "a dime is a round silver thing" or "a circular metal object." Now, begin listing descriptive features, as volunteered by the class, on the board. List everything offered. Then, after asking the class to identify the pertinent features, compose drafts of the description on the board. Working together, the class should eventually arrive at a version something like this:

> A dime is a tri-layered, bimetallic disk, 11/16 inch (17 mm) in diameter, 1/20 inch (1.5 mm) thick, weighing roughly 0.07 ounce (2 g). The center layer of copper is bonded between two nickel surfaces, bordered by a 0.2-mm raised, rolled rim with a perimeter of equally spaced serrations at 0.2-mm intervals, perpendicular to the flat surfaces (or parallel to the vertical axis).

Of course, how something is described depends on the writer's purpose and the audience's needs. Technical writers name things in ways that have significance and are useful to a specific audience.

In any performance course, students want to know immediately what is expected of them. As a general summation of the syllabus, Chapter 1, and the course description, you might tell students that success in the course depends on their meeting three general requirements: (1) attending class regularly and participating actively; (2) following directions; and (3) meeting all deadlines. Refer to these requirements throughout the semester.

To make your approach concrete at once, you might use exercises 1 and 4 in sequence. Complement exercise 1 with the plus-or-minus column analogy on Master Sheet 8. After the visual presentation and discussion of samples, students have enough background to handle the persuasive-writing situation in exercise 4 (or in exercise 5, for students who work full time).

Master Sheets 9 through 16 illustrate the writing process as a set of critical-thinking decisions that are deliberate rather than random, and recursive rather than linear. You might use this material early in the course or as a supplement to Chapter 7, or at both stages.

Master Sheet 7

Guidelines for Memo Formatting

NAME OF ORGANIZATION

MEMORANDUM
Date: (also serves as a chronological record for future reference)
To: Name and title (the title also serves as a record for reference)
From: Your name and title (your initials for verification)
Subject: GUIDELINES FOR FORMATTING MEMOS

Subject Line
Announce the memo's purpose and contents, to orient readers to the subject and help them assess its importance. An explicit title also makes filing by subject easier.

Introductory Paragraph
Unless you have reason for being indirect, state your main point immediately.

Topic Headings
When discussing a number of subtopics, include headings (as shown here). Headings help you organize, and they help readers locate information quickly.

Visuals
To summarize numerical data, your memo may include one or more visuals.

Paragraph Spacing
Do not indent the first line of paragraphs. Single space within paragraphs, and double space between them.

Second-page Header
When the memo exceeds one page, include a header on each subsequent page. For example: Ms. Baxter, June 12, 20xx, page 2.

Memo Verification
Do not sign your memos. Initial the "From" line, after your name.

Copy Notation
When sending copies to people not listed on the "To" line, include a copy notation two spaces below the last line, and list, by rank, the names and titles of those recipients. For example,

Copies: J. Spring, V.P., Production
         H. Baxter, General Manager, Production

Master Sheet 8

How a Document Is Evaluated

Deciding how well a document *communicates,* users place it (and its author) immediately in the *Plus* or *Minus* column:

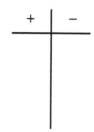

Specifically, users evaluate your message by applying these four general questions:

- Is the document appealing?
- Is the information worthwhile?
- Is the message easy to follow?
- Is the message easy to read?

The answer to each of the above questions should be *Yes:*

Otherwise, your message fails. Even one *Minus* feature can erase the remaining *Plus* features:

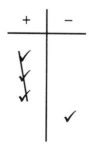

A document is evaluated by the quality of its *appearance, content, organization,* and *style.*

Master Sheet 9

How a Document Is Composed

All effective writing is a thinking and problem-solving process.

Writing is recursive. That is, as we write, we move back and forth between and among these steps: planning, drafting, and revising until we reach our goal of crafting a useful message. In this example, the arrows illustrate some of the forward and backward movements we might make during the process of writing.

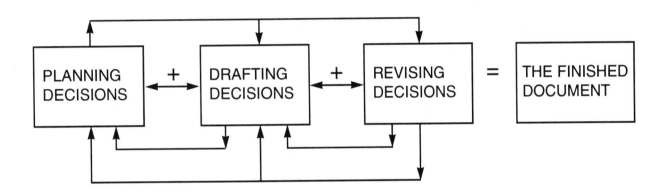

As we move back and forth—planning, drafting, rethinking, and revising—we're not merely recording information. Instead, we are discovering and selecting suitable content, organizing it, and refining it.

In essence, during this problem-solving process, we're making hundreds of deliberate decisions about our purpose and our audience's needs. (Some decisions might evolve after hours of thinking and planning; others are nearly instantaneous.)

These countless decisions about content, arrangement and style collectively make up the *writing process.*

Master Sheet 10

The Writing Process for Technical Documents

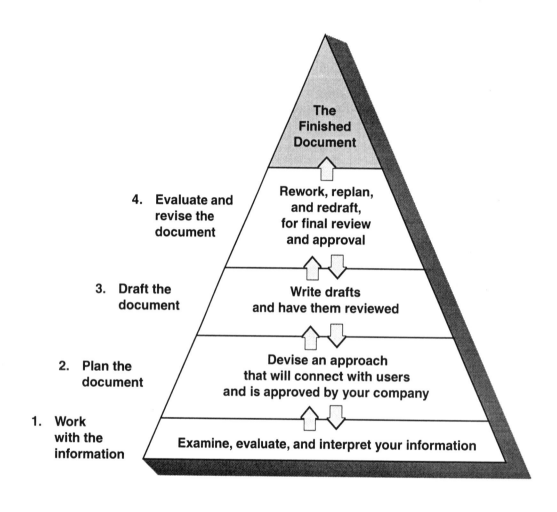

The Finished Document

4. **Evaluate and revise the document** — Rework, replan, and redraft, for final review and approval

3. **Draft the document** — Write drafts and have them reviewed

2. **Plan the document** — Devise an approach that will connect with users and is approved by your company

1. **Work with the information** — Examine, evaluate, and interpret your information

Master Sheet 11

Creative and Critical Thinking in the Writing Process

1. **Work with the information:**
   - ☐ Have I defined the problem accurately?
   - ☐ Is the information complete, accurate, reliable, and unbiased?
   - ☐ Can it be verified?
   - ☐ How much of it is useful?
   - ☐ Do I need more information?
   - ☐ What do these facts mean?
   - ☐ What connections seem to emerge?
   - ☐ Do the facts conflict?
   - ☐ Are other interpretations possible?
   - ☐ Is a balance of viewpoints represented?
   - ☐ What, if anything, should be done?
   - ☐ Is it honest and fair?
   - ☐ Is there a better way?
   - ☐ What are the risks and benefits?
   - ☐ What other consequences might this have?
   - ☐ Should I reconsider?

2. **Plan the document:**
   - ☐ When is it due?
   - ☐ What do I want it to do?
   - ☐ Who is my audience, and why will they use it?
   - ☐ What do they need to know?
   - ☐ What are the "political realities" (feelings, egos, cultural differences, etc.)?
   - ☐ How will I organize?
   - ☐ What format and visuals should I use?
   - ☐ Whose help will I need?

3. **Draft the document:**
   - ☐ How do I begin, and what comes next?
   - ☐ How much is enough?
   - ☐ What can I leave out?
   - ☐ Am I forgetting anything?
   - ☐ How will I end?
   - ☐ Who needs to review my drafts?

4. **Evaluate and revise the document:**
   - ☐ How does this draft measure up?
   - ☐ Does it do what I want it to do?
   - ☐ Is the content worthwhile?
   - ☐ Is the organization sensible?
   - ☐ Is the style readable?
   - ☐ Is everything easy to find?
   - ☐ Does the format look good?
   - ☐ Is everything accurate, complete, appropriate, and correct?
   - ☐ Who needs to review and approve the final version?
   - ☐ Does it advance my organization's goals?
   - ☐ Does it advance my audience's goals?

Master Sheet 12

A Flowchart of the Writing Process

Decisions in the writing process are recursive: no one stage of decisions is complete until *all* stages are complete.

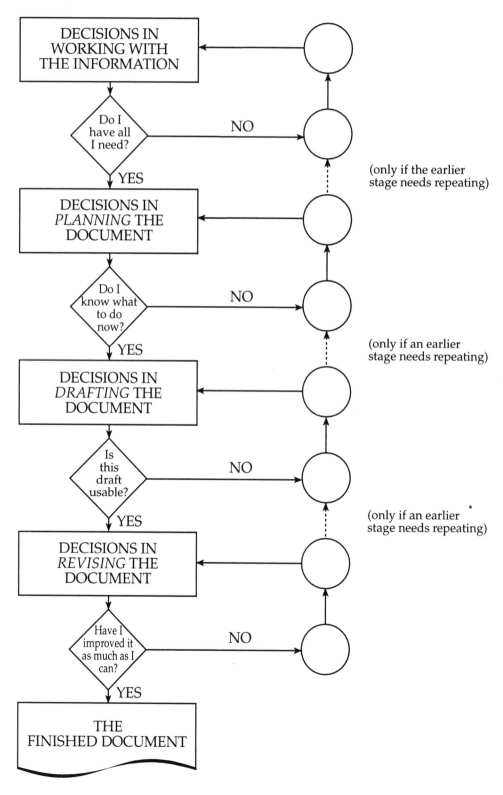

Master Sheet 13

Decisions in Planning the Document

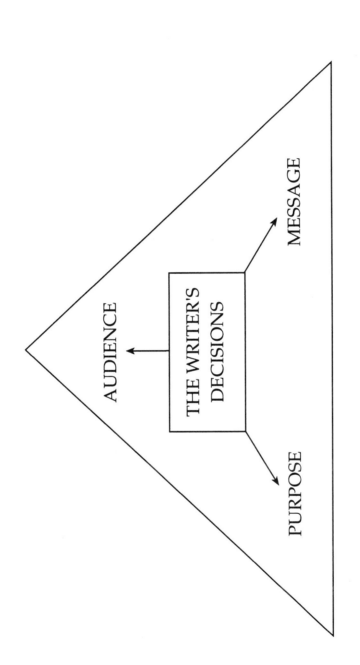

Purpose

- Why and when is this document needed?
- What do I want it to achieve (describe, explain, persuade, advise, recommend, evaluate)?

Audience

- Who are my (primary and secondary) audiences, and how specialized are they?
- What questions do they need answered?

Message

- What meaning do I want to convey?
- What material will I need, and where will I get it?

Master Sheet 14

Decisions in Drafting the Document

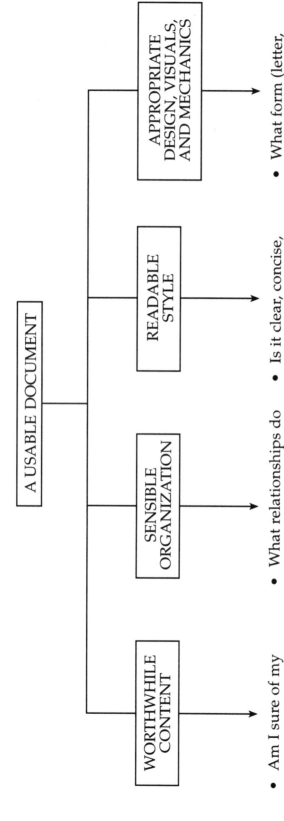

A USABLE DOCUMENT

WORTHWHILE CONTENT

- Am I sure of my meaning?
- Do I have enough material?
- Have I thought enough about my material?
- Which material is relevant?
- How much is enough?
- Can I eliminate anything?
- Have I forgotten anything?

SENSIBLE ORGANIZATION

- What relationships do these data suggest?
- What do I want to emphasize?
- In which sequence will users approach this material?
- What belongs where?
- What do I say first?
- What comes next?
- How do I end?

READABLE STYLE

- Is it clear, concise, and fluent?
- Is it too technical or too simple?
- Am I writing as if I were speaking?
- What tone and attitude does my audience expect?

APPROPRIATE DESIGN, VISUALS, AND MECHANICS

- What form (letter, memo, report) should I use?
- Do I need supplements (abstract, appendixes, other)?
- How should I use headings and white space?
- Can I use visuals?
- Do I need to check my dictionary or handbook?

Master Sheet 15

Decisions in Revising the Document

Is the Content Worthwhile?

- ☐ A brief but explicit title

- ☐ Subject and purpose clearly stated

- ☐ Enough information for readers to understand the meaning

- ☐ Material (or insight) new and significant to the audience

- ☐ All material technically accurate

- ☐ Technical details appropriate for the audience

- ☐ All needed warnings and cautions

- ☐ All data examined fully and interpreted impartially

- ☐ Both sides of the issue presented

- ☐ Opinions and assertions supported by evidence

- ☐ Conclusions and recommendations supported by the facts presented

- ☐ No recommendations where none were requested

- ☐ No gaps, foggy areas, or needless details

- ☐ All anticipated reader questions answered

- ☐ All data sources documented

Is the Organization Sensible?

- ☐ Structure of the document visible at a glance

- ☐ An evident line of reasoning

- ☐ A distinct introduction, body, and conclusion

- ☐ A section's length that is equal to its importance

- ☐ Enough transitions and connectors to signal relationships

- ☐ Material organized for the user's understanding

- ☐ A topic (orienting) sentence to begin each supporting paragraph

- ☐ One main point developed in each supporting paragraph, with unity, coherence, and reasonable length

Master Sheet 16

Decisions in Revising the Document

Is the Style Readable?

□ Each sentence understandable on *first* reading (clarity)

□ Most information expressed in fewest words (conciseness)

□ Related ideas combined for fluency

□ Sentences varied in construction and length

□ Each word chosen for exactness, not for camouflage

□ All definitions double-checked

□ Concrete and specific language

□ No triteness, overstatements, euphemisms, or inappropriate jargon

□ Tone unbiased and inoffensive

□ Level of formality appropriate to the situation

Are Design, Visuals, and Mechanics Appropriate?

□ An inviting and accessible format: white space, fonts, and so on

□ A design that accommodates audience needs and expectations

□ Adequate, clear, and informative headings

□ Adequate visuals, to clarify, emphasize, or organize

□ Appropriate displays for specific visual purposes

□ All visuals fully incorporated with the text

□ All visuals free of distortion

□ All pages numbered and in order

□ Supplements that accommodate diverse audience needs

□ Correct spelling, punctuation, and grammar

Master Sheet 17

Chapter 1 Quiz

Name _____ Section _____

Indicate whether statements 1–5 are TRUE or FALSE by writing *T* or *F* in the blank.

1. _____ In the best technical documents, the writer "disappears."

2. _____ Raw data is an important type of information.

3. _____ Effective communicators "let the facts speak for themselves."

4. _____ As you advance in your profession, your ability to communicate is likely to become more important than your technical background.

5. _____ Direct, straightforward communication is valued by all cultures.

Complete statements 6–7.

6. A computer can transmit data, but it cannot give_____ to the information.

7. Executives consistently rank_____ skills as the most vital of employee qualities.

In 8–10, choose the letter of the expression that best completes each statement.

8. _____ A technical document focuses on (a) the subject, (b) the writer's feelings, (c) both the subject and the writer's feelings, (d) marketing, or (e) none of these.

9. _____ A technical document is based on (a) intuition, (b) specialized information, (c) the writer's deepest impressions, (d) inspiration, or (e) none of these.

10. _____ The language of a technical document is (a) entertaining, (b) efficient, (c) confidently judgmental, (d) prosaic, or (e) none of these.

PART

I

Communicating
in the Workplace

This section creates a problem-solving context for the writing challenges in later chapters. Besides offering a rationale for the course—an answer to "Why are we doing this?"—Part I promotes audience awareness and critical-thinking skills. Students learn to think critically about the informative, persuasive, ethical, global, and collaborative dimensions of their communications.

2

Problem Solving in Workplace Communication

Chapter 2 introduces students to the reality of communicating in a workplace in which the rhetorical complexities transcend linear notions of merely "transmitting" information.

The intent here is not to overwhelm students, but to help them define the kinds of problems they need to solve, the range of decisions they need to make, and the types of strategies they might employ for effective decision making.

Master Sheets 18 and 19 offer an introduction to audience awareness.

Master Sheet 18

Sample Situations for Communicators

In many jobs you will have to ask questions daily about your audience's needs. Here are two scenarios in which specialists who also must be "part-time" technical communicators assess their audiences' needs in different situations.

A Day with a Police Officer

You are a police officer in the burglary division of an urban precinct.

- A local community college has asked you to lecture on the fields of police work during its career week. Your audience will be interested laypersons, some considering a police career. You know they've been captive audiences at lectures for years, and so you decide to liven things up with visuals. The police artist agrees to draw up a brightly colored flip chart and you use graphics software on the office microcomputer to generate graphs and flowcharts illustrating salaries and career paths.

- Tomorrow in court you will testify for the prosecution of a felon you caught red-handed last month. His lawyer is a sly character, known for making police testimony look foolish, and so you are busy reviewing your notes and getting your report in final form. Your audience will consist of legal experts (judge, lawyers) as well as laypersons (jury). The key to a convincing testimony will be *clarity, accuracy,* and *precision,* with impartial and factual descriptions of *what* happened, as well as *when* and *where.*

- You are drafting an article about new fingerprinting techniques for a law enforcement journal. Your readers will be expert (veteran officers) and informed (junior officers). Because they understand shoptalk and know the theory and practice of fingerprinting, they won't need extensive background or definitions of specialized words. Instead, they will read your article to learn about a new *procedure.* You will have to convince them that your way is better. You will later write a manual for using this technique.

- Your chief asks you to write a manual on investigative procedures for junior officers in your division. The manual will be bound in pocket size and carried by all junior officers on duty. Your audience is informed, but not expert. They have studied the theory at the police academy, but now they need the "how-to." They will want rapid access to instructions for handling problems, with each step spelled out. They probably will use your instructions as a guide to *immediate* actions and responses. Therefore, you will have to label warnings and cautions *before* each step. Clarity throughout will be imperative.

- Next month you will speak before the chamber of commerce on "Protecting Your Business Against Burglary." The audience will be highly interested laypersons. Because burglary protection can be costly (alarm systems, guard dogs), the audience is apt to be most interested in the less expensive precautions they can take. They will want to remember your advice, and so you decide to supplement your talk with visuals: specifically, a checklist for business owners that you will discuss item-by-item, using an overhead projector.

Master Sheet 19

A Day with a Civil Engineer

As a civil engineer in a materials-testing lab, you are in charge of friction studies on road surfaces.

- Your firm is competing for a contract to study the safety of state bridge surfaces. As engineer in charge, you are responsible for the quality of this proposal. Thus you are carefully reviewing and editing, online, proposal sections written by employees. Your audience consists of experts (state engineers), informed readers (officers in the highway department), and laypersons (members of the state legislature). The legislature ultimately will decide which firm gets the contract, but they will act on the advice of engineers and the highway department. You decide therefore to keep your proposal at a level of technicality that will connect with the specialized audience. The legislative committee will read only the informative abstract, conclusions, and cost data.

- Checking your electronic mail, you find a message from the vice president asking you to write safety instructions for paving crews on the new bridge. Your instructions will be read by all crew members, from project engineer to laborer, but when it comes to safety on this kind of job, even the specialized personnel may be considered laypersons. You ask the graphics department to design a brochure for your instructions, including sketches and photographs of the hazardous areas and situations.

- At next month's convention in Dallas you are scheduled to deliver a paper describing a new road-footing technique that reduces frost heave damage in northern climates. (This paper will soon be published in an engineering journal.) Your audience will be civil engineering colleagues who want to know what it is and how it works—without lengthy background. You decide to compress your data with conventional visuals (cross-sectional drawings, charts, graphs, maps, and formulas), as well as computer graphics.

- You draft a letter to a Korean colleague you've never met, to ask about her new technique for increasing the durability and elasticity of rubberized asphalt surfaces. Your expert reader is a busy professional with little time to read someone's life story, but she will need *some* background about your own work, along with clear, specific, and precise questions that she can answer quickly. As a gesture of goodwill, you close your letter with an offer to share *your* findings with her.

- You draft and revise a memo justifying your request to the vice president for an additional lab technician. You know that the present lab technician feels threatened by the prospect of someone invading his space, and so you have to be sensitive, diplomatic, and persuasive—and to make sure that Joe (the lab technician) receives a copy of your memo.

Master Sheet 20

Chapter 2 Quiz

Name _____ Section _____

Indicate whether statements 1–6 are TRUE or FALSE by writing *T* or *F* in the blank.

1. _____ As long as they know the facts, audiences can interpret them easily.

2. _____ Effective communication ensures that the interests of your company take priority over the interests of outside audiences.

3. _____ Automated grammar and style "checkers" do not eliminate the need for careful proofreading.

4. _____ Research shows that people in general trust printed text more than visual images.

5. _____ The writing process has four stages.

6. _____ Each stage of the writing process must be completed before the next stage can be attempted.

In 7–10, choose the letter of the expression that best completes each statement.

7. _____ Technical communicators encounter all of the following problems except (a) the information problem, (b) the persuasion problem (c) the confabulation problem, (d) the ethics problem, and (e) the collaboration problem.

8. _____ Workplace problem solving requires (a) creative thinking, (b) critical thinking, (c) transitive thinking, (d) b and c, or (e) a and b.

9. _____ Solving the persuasion problem means (a) using whatever works, (b) building a reasonable case, (c) appealing to emotion, (d) all of these, or (e) none of these.

10. _____ Critical thinking (a) is derived from literary criticism, (b) is used only in emergencies, (c) involves weighing alternatives, (d) both a and b, or (e) a, b, and c.

CHAPTER
3

Solving the Information Problem

Analyzing the audience is one of the most important (and elusive) skills students can develop. In the workplace and in school, inexperienced writers often are unaware of the need to adapt a message to their audience. In their simplistic view, writing is a linear task of transferring material from the brain to the page. Without a sense of their audience, writers write prematurely—and thus ineffectively.

Spend some time on the Levels of Technicality section (pages 23–28), analyzing each sample to see how the level of technicality is adjusted to the audience's expectations and needs. Students with traditional composition backgrounds need practice in thinking about their readers' specific needs for clear and useful information.

Tell students you will read and evaluate their writing as an employer or supervisor would—a decision maker who requires clear information, often translated from high to low technicality. (Here is where contract grading fits in: in the workplace, a product is unacceptable, acceptable, or superior.) Have students identify an audience and use for each assignment. You might want them to include a written audience-and-use analysis with each submission—especially for the earlier assignments.

If you are unfamiliar with a particular specialty (such as computer science or electrical engineering), ask students planning long reports or proposals about these specialties to use you as the *secondary* reader, and to prepare the report text and supplements accordingly. For class discussion, ask students to describe situations in which they've had to explain something specialized to an uninformed audience (such as camp counselors, hobbyists, part-time employees). Or ask them to describe situations in which school lectures have sailed over their heads, and to analyze the reasons.

Students invariably ask, "How long should this assignment be?" as they try to apply versions of the "500-word essay" formula to all assignments. My response: "Just long enough to answer all anticipated questions from the audience." Explain that writers who can accurately anticipate their audience's questions are those who know how much is enough. Discuss briefly the audience-and-use profiles preceding sample reports (pages 432, 434, 449, 478, 509, 540) to show how writers adjust their level of detail to audiences.

During editing workshops throughout the semester, emphasize repeatedly that every word, sentence, and paragraph should advance the writer's meaning. Chapter 13 provides basic editing tools for achieving clear and precise expression.

Exercise 1 works well for students with some technical sophistication. Emphasize that the workplace communicator writes for audiences who know less about the subject than the writer (as opposed to writing for professors, who know more about the subject than the student writer). When the specialized student writes for you and heterogeneous classmates, he or she becomes the teacher and the readers become the students. Given this context, students see writing as more than throwing words down on the page as they are peeled off the top of one's head; instead, they see writing as a set of decisions based on careful consideration of subject, situation, and audience. When writers connect with their audience, they succeed; when they don't, they fail.

Master Sheet 21 offers another twist for connecting with an audience. You might want to ditto this material as a handout, and refer to it when the class works on complaint and job-application letters and justification reports.

The Chapter 3 Collaborative Project is essential for students at any level. This project helps develop audience awareness by guiding students through their own detailed analysis of their audience's needs, attitudes, and expectations.

Use Master Sheets 23–25 (on the opaque projector or as transparencies) to enhance class discussion in preparation for this team project. Master Sheets 26 and 27 show one possible set of responses for the audience analysis in the collaborative assignment. (For general suggestions about assigning collaborative projects, see page 73 in this manual.)

Master Sheet 21

Coping With a "Dangerous" Audience

It would be naive and misleading to suggest that just thinking about your audience will help solve *all* your writing problems. Thinking about some audiences, in fact, can so intimidate writers that they "choke." This block is especially common when you are reporting bad or surprising news, making a complaint or an unpopular suggestion to superiors, or when much attention is to be focused on your report. Often, the instructor-as-audience can be intimidating as well.

If for any reason you think your audience might be unreceptive, or "dangerous," try writing the first draft for yourself or for a different audience. Writing specialist Peter Elbow suggests: "For example, you can address a draft of your technical report to your loved one—even permitting yourself some of the fun and games your make-believe audience inspires."[1] By imagining a different audience (or none at all), you can sometimes discover clearly where you stand *before* trying to connect with your real audience. Once you've discovered what to say and how to say it, adjusting the message to your real audience is easy.

---

[1]For some excellent strategies in coping with a "dangerous" audience, see Elbow, Peter. *Writing with Power* (New York: Oxford, 1981: 187–190).

Master Sheet 22

Deciding on a Document's Level of Technicality

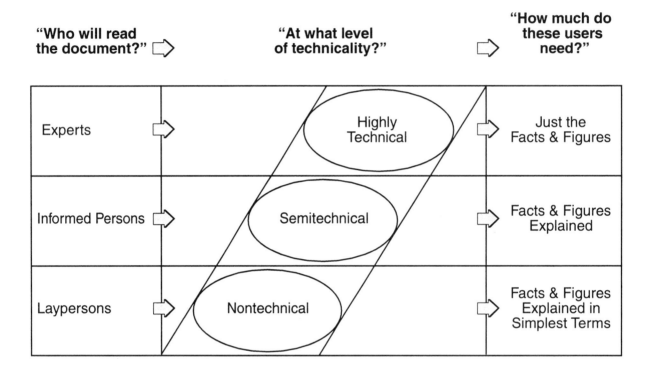

Master Sheet 23
Audience-And-Use Profile Sheet

Learn all you can about the audience before you communicate.

Identity and Needs

1. Who is my primary audience? Who else will read the document?
2. What is my relationship to this audience?
3. How will my document be used (to solve a problem, make a decision, perform a task, answer a question)?
4. How much is my audience likely to know already about this topic?
5. What else does the audience need to know (background, definition, and so on)?
6. What main questions are most users likely to have?

Attitude and Personality

7. What attitudes or misconceptions is the audience likely to have toward the topic? Are they likely to have any objections?
8. What attitude does the audience seem to have toward me?
9. How receptive to new ideas is this organization?
10. Who will be most affected by this document?
11. What do I know about the user's (or group's) personality (way of thinking, behaving, and reacting)?
12. What reaction to this document can I expect?
13. Do I risk alienating anyone?
14. Do I face any constraints?

Expectations About the Document

15. Has this document been requested or am I initiating it?
16. How will the cultural context shape this audience's expectations?
17. What length will the audience expect and tolerate (spell it out or keep it short and sweet)?
18. For this audience, what kinds of details will be most important (conclusions, a summary, cost factors, how the material affects them)?
19. How would they expect the piece to be organized?
20. What tone would this audience expect?
21. What is this document's intended effect on its audience?
22. When is the document due?

Master Sheet 24

Audience-And-Use Profile Sheet

Identity and Needs

- My primary audience is _____.
  (name, title)

- Other potential users are _____.

- The audience is related to me as a _____.
  (client, employer, employee, supervisor, colleague, friend, close acquaintance, distant acquaintance, stranger, person doing a favor, person receiving a complaint, other)

- The audience will use my document to _____.
  (solve a problem, make a decision, answer a question, take an action, carry out a procedure, improve performance, take a stand on some issue, learn about something new, receive good or bad news, other)

- The audience probably knows_____
  about this topic. (nothing, very little, the general background, quite a few details)

- The audience still needs _____
  in order to understand this document. (definitions, background, item-by-item explanation, a summary, only the bare facts, interpretations and conclusions spelled out, other)

- The audience is likely to have these important questions:

  _____?

  _____?

  _____?

  _____?

Attitude and Personality

- In its attitude toward this topic, the audience is likely to be _____.
  (indifferent, biased, misinformed, defensive, skeptical, interested, uncertain, confused, other)

- Audience objections are likely to include _____.
  (cost, labor, time, fear of consequences, none, other)

Master Sheet 25

Audience-And-Use Profile Sheet

- In its attitude toward me, the audience seems to feel _____.
  (*intimidated, superior, hostile, receptive, indifferent, unsure, threatened, confident, other*)
- The organizational climate seems _____.
  (*competitive, repressive, cooperative, creative, resistant to change, other*)
- Those most affected by this document will be _____.
  (*primary audience, secondary audience, persons who have not read the document, other*)
- This audience's temperament in this situation seems likely to be_____.
  (*domineering, short-tempered, cautious, impatient, impulsive, supportive, demanding, tolerant, analytical, insecure, other*)
- I can expect the audience to react with_____.
  (*confusion, fear, guilt, resistance, shock, anger, annoyance, resentment, approval, appreciation, other*)
- People I might alienate with this document are _____.
  (*colleagues, superiors, subordinates, clients, other*)
- Constraints I face include _____.
  (*time, legal, ethical, other*)

Expectations about the Document

- This document is being written_____.
  (*at the audience's request, on my initiative, other*)
- The cultural context may cause this audience to focus on _____.
  (*the importance of indirectness, face-saving, other*)
- The kinds of information that will be most important to this audience are_____
  _____.
  (*interpretations, conclusions, recommendations, a summary, costs, expected results, benefits, descriptive or procedural details, other*)
- The audience would expect the document to be organized in this way: _____
  _____.
  (*problem-causes-solution, questions-answers-conclusions-recommendations, reasons for/reasons against, proposed action-probable effects-conclusions, item-by-item or point-by-point comparison, other*)
- This audience would expect a(n) _____ tone.
  (*formal, informal, conversational, relaxed and friendly, serious and businesslike, enthusiastic, impartial, apologetic, indignant, other*)
- This document's intended effect on its audience is to _____.
  (*win the audience's support for a project, position, or idea; bring about a definite action; change behavior, instruct about a procedure; keep the audience informed; retain the audience's goodwill; other*)

Master Sheet 26

A Sample Audience Analysis for the Chapter 3 Collaborative Assignment

- *Who is my audience?* Incoming students in the major (and faculty).

- *How will this audience use the information?* To develop a sense of what to expect and how to proceed (say, in managing workloads or meeting deadlines).

- *How much is the audience likely to know about this topic?* Very little. They need everything spelled out.

- *What else does the audience need to know?* They need answers to questions like these: How big is the problem? How can it affect me? What are the department's expectations? How much homework will I need to do? How should I budget my time? Can I squeeze in a part-time job? Are there any skills I should try to acquire beforehand (say, word processing or graphics and basic design skills)? Where do most first-year students make their big mistakes? (Based on your own experience, can your team anticipate any other questions?)

- *What attitudes or misconceptions about this topic is the audience likely to have?*

  Any who are overly optimistic ("No problem!") will need to visualize the real challenges ahead.

  Any who are overly pessimistic ("I'm dead for sure!") will need encouragement, along with the facts.

  Any who are indifferent ("Who cares?") will need some motivation, along with the facts.

  Whatever combination of attitudes the audience holds, we have to address each attitude—as well as we can identify it.

- *What probable attitude does the audience have toward the writers?* (Are we seen as trustworthy, sincere, threatening, arrogant, or what?) Since we are all students, readers will likely identify with and trust us to an extent. They'll probably realize we're on their side.

Master Sheet 27

A Sample Audience Analysis for the Chapter 3
Collaborative Assignment (continued)

- *Who will be affected by this document?* Mostly the incoming students (primary audience), and possibly the department.

- *In this situation, how can we characterize the audience's temperament and probable reaction?* Most readers should be eager for this information and should take it seriously.

- *Do we risk alienating anyone?* Gifted students who don't know the meaning of failure might feel patronized or offended. Some faculty might resent any suggestions that courses are too demanding, and so we don't want to editorialize. The purpose of this piece is informative and advisory—not evaluative.

- *How did this document originate, and how long should it be?* Because it was requested by the department and not by the primary audience, we can't expect students to tolerate more than a page or two.

- *What material will be most important to this audience?* They will want clear advice about what and what not to do.

- *What arrangement would be most effective for this audience and purpose?* We should provide brief background on the dropout problem, discuss its causes, suggest ways to survive, and end on a positive note of encouragement and motivation.

- *What tone would this audience expect?* We are all students; a friendly, relaxed, and positive (to avoid panic), but serious tone seems best.

- *What is the document's intended effect on its audience?* If it manages to connect it will, we hope, cut down the dropout rate.

Master Sheet 28

Chapter 3 Quiz

Name _____ Section _____

Indicate whether statements 1–6 are TRUE or FALSE by writing *T* or *F* in the blank.

1. _____ Primary and secondary audiences often have different technical backgrounds.

2. _____ When unable to identify all members of an audience, you should aim at the least specialized members.

3. _____ Primary audiences usually expect a semitechnical message.

4. _____ Information needs may be culturally determined.

5. _____ Audience analysis is only necessary when the document is long or complex.

6. _____ Nontechnical audiences merely are interested in the bare facts, without explanations.

Respond to 7–10.

7. We focus on our audience and purpose by answering these questions: Who wants the report? Who else will read it? List two other questions writers ask about their audience.

8. Briefly explain the difference between writing for a college professor and writing for an audience on the job.

9. When does a message have informative value for its audience?

10. Briefly explain the difference between primary and secondary audiences.

CHAPTER
4

Solving the Persuasion Problem

In response to calls for expanded treatment of persuasion in professional writing texts, this chapter is based on the latest findings in rhetoric, social science, and communication theory. Treated here are interpersonal problems routinely confronted by writers in the workplace. Students need to understand that any piece of writing can be redefined by each reader—depending on that reader's biases, preferences, motives, or attitude. To introduce the notion of interpersonal problem solving, I usually say something like this:

> Your suggestions or ideas might impress one reader while enraging or offending someone else. Your major task as a writer is to do everything you can to ensure that your document has the effect *you intend* on your audience. Even the clearest and most informative communication can spell disaster if a writer has ignored the situation's political realities.

Effective communicators are effective critical thinkers; they know how to ask for things, or how to instruct or warn or direct or advise. They know how to avoid asking for too much, how to respect a situation's constraints, and how to support their claims convincingly.

Technical/professional writing has an immediate and measurable *effect* on the audience— and, in turn, on the writer.

Along with Chapter 5, "The Ethics Problem," Chapter 4 views workplace communication as a set of rhetorical problems involving more than mere "information transport."

Master Sheet 29 offers an authoritative summary of persuasive skills needed by engineers.

Master Sheets 23–25 from Chapter 3 can be used here as well, for discussing audience-and-use analysis.

Master Sheets 30 and 31 show one possible set of responses for Goal "e" in the second collaborative assignment.

Master Sheet 29

Persuasion Pays

[Today's] engineer . . . must be a Renaissance person to succeed, says Lawrence J. Kamm, a veteran San Diego consulting engineer with 47 years of "engineering and entrepreneuring" experience and the author of . . . *Successful Engineering: A Guide to Achieving Your Goals*. Kamm's rules for success for entry-level engineers include an analytical career plan for the long term, and understanding of engineering economics, art, and history, and a knowledge of the techniques and problems of career management.

"Young engineers fresh from school believe that if they do good work its merit will persuade others to use it," says Kamm. "They are absolutely right. The most persuasive of all the arguments in selling engineering ideas and work is their intrinsic value."

"Yet writing is the bane of engineers; we are almost proud to do it badly. Yet we must do some of our persuading in letters, reports, and proposals. Good writing is like good design development; you must repeatedly refine your work."

In short, tomorrow's engineer cannot afford to be one-dimensional. "The most valued skill an engineer has," according to Kamm, "the thing that brings the engineer the greatest income and satisfaction, and job security, if there is such a thing, comes from the power of persuasion."

"All technical knowledge is a subclass," he says, "and not particularly persuasive." But persuasion on the job becomes important because engineers deal daily with budgets, requesting more money for projects, asking for raises, delegating jobs to subordinates, and selling superiors on changes in design and myriad other things that can affect the engineer's company, job, and career.

"Technical engineering," says Kamm, "is one technique of persuasion because of an engineer's knowledge of the subject. The ability of the engineer to express himself or herself in clear and forceful English is another technique of persuasion without which the first one doesn't get very far."

Engineers, says Kamm, are notorious for ignoring their interpersonal skills. " 'Twas ever thus," says Kamm. "But those who do learn the valuable art of persuasion rise highest in management and do not remain as technical people all their lives."

The lesson learned is that the well-rounded engineer would be well advised to cultivate the powers of persuasion to ensure success in this decade. In short, persuasion pays, as all good persuaders know.

—*J. Robert Connor, Editor-in-Chief*

Master Sheet 30

A Sample Audience Analysis for the Chapter 4 Collaborative Assignment (Goal "e")

- *Who is my audience?* Classmates and instructor (and possibly some college administrators).

- *How will my information be used?* People will decide whether to support our recommendation for limiting possible grades in this course to three: A, B, or F.

- *How much is the audience likely to know already about this topic?* Everyone here is already a grade expert, and will need no explanation of the present grading system.

- *What else does the audience need to know?* The instructor should need no persuading; he or she knows all about the quality of writing expected in the workplace. But some of our classmates probably will have questions like these: Why should we have to meet such high expectations? How can this grading be fair to the marginal writers? How will I benefit from these tougher requirements? Don't we already have enough work here?

  We will have to answer questions by explaining how the issue boils down to "suffering now" or "suffering later," and that one's skill in communication will determine one's career advancement.

- *What attitude or misconceptions about the topic is our audience likely to have?*

  If they are indifferent ("Who cares?"), we have to encourage them to care by helping them understand the eventual career benefits of our recommended grading system.

  If they are interested ("Tell me more!"), we have to hold their interest and gain their support.

  If they are skeptical ("How could this plan ever work?"), we have to show concretely how the plan could succeed.

  If they are biased ("I hate writing."), we have to emphasize how greatly writing matters in the workplace.

  If they are misinformed ("I'll have secretaries or word processors to fix up my writing."), we have to provide the correct information, backed by solid evidence.

  If they are defensive ("Why pick on me?"), we have to persuade them that our idea is constructive, that we are genuinely supportive, and that we have their best interests in mind.

  If they have realistic objections ("Weaker writers will be unfairly penalized," or "Some students can't afford more than one semester for this course."), we need to offer a plan addressing these objections, and to show how the benefits can outweigh the objections.

  Whichever combination of attitudes our audience holds, we have to do our best to satisfy each reader's objections—as well as we can identify them. And if some readers are dead set against the idea, we can't expect to convert them, but we can encourage them at least to consider our position.

Master Sheet 31

A Sample Audience Analysis for the Chapter 4 Collaborative Assignment (continued)

- *What is the audience's attitude toward the writer(s), before anyone has read the document? (Is the writer seen as trustworthy, sincere, threatening, arrogant, meek, underhanded, or what?)* We need to gain our audience's confidence, to make sure they interpret our motives not as totally self-serving or elitist, but as sincere and caring about the welfare of the whole group.

- *What is the organizational climate? (Is it competitive, repressive, cooperative, creative, resistant to change, or what?)* In this class, we can all feel comfortable about speaking out.

- *Who will be most affected by this document?* The primary audience, our classmates.

- *In this situation, how can we characterize the audience's temperament and probable reaction?* Many classmates are likely to feel threatened, and probably will react initially with resentment and resistance.

- *Do we risk alienating anyone?* Yes, especially students whose writing never has earned a grade of B or better.

- *How did this document originate, and how long should it be?* Because we initiated the document, we can't expect readers to tolerate a long, involved presentation. To be persuasive, however, we do have to make our case concrete.

- *What material will be most important to this audience?* They will want a clear picture of the benefits in such an apparently radical plan.

- *What arrangement would be most effective for this audience and purpose?* We should propose the new grading system, offer our reasons, point out the benefits, show how the system could operate, and close with a request for support.

- *What tone would this audience expect?* We are all at least acquainted; a friendly, conversational tone seems best.

- *What is this document's intended effect on its audience?* If it manages to connect with its audience, this document will win their support for our recommendation.

Follow the model in Figure 4.6 for designing a profile sheet to record your audience-and-use analysis, and to duplicate for use throughout the semester. (Feel free to improve on the design and content of our model.)

Master Sheet 32

Chapter 4 Quiz

Name_____Section _____

Indicate whether statements 1–6 are TRUE or FALSE by writing *T* or *F* in the blank.

1. _____ An effective technical document never allows its audience to read anything between the lines.

2. _____ Co-workers tend to be easily persuaded.

3. _____ The persuader's likability often is the biggest factor in persuasion.

4. _____ Audiences create their own meanings from what they read.

5. _____ People almost always are persuaded by sound reasoning.

6. _____ A persuasive message begins with a request or claim that recipients will reject, and then asks for something more reasonable.

Choose the letter of the expression that best completes each statement.

7. _____ An audience ideally responds to persuasion through (a) compliance, (b) internalization, (c) obfuscation, (d) elevation, or (e) cogitation.

8. _____ The longest-lasting connection between persuader and audience tends to be (a) the rational connection, (b) the relationship connection, (c) the power connection, (d) the time/space connection, or (e) the love connection.

9. _____ All of the following are communication constraints except (a) legal constraint, (b) ethical constraint, (c) time constraint, (d) transportation constraint, or (e) social constraint.

10. _____ Convincing evidence includes everything except (a) statistics, (b) examples, (c) speculation, (d) expert testimony, or (e) all information that supports your claim.

CHAPTER
5

Solving the Ethics Problem

This chapter introduces the long-overlooked notion of *accountability*, and serves as a reference point for ethical considerations throughout the text and the course. Students need to understand that their communications choices have definite ethical consequences, and that standards of usefulness and persuasiveness have as corollaries standards of honesty and fairness.

Chapter 5 further expands our definition of the communication problem faced by workplace writers:

1. "How do I give readers the information they need?" (The Information Problem)

2. "How can I get the response I want?" (The Persuasion Problem)

3. "How can I do the right thing?" (The Ethics Problem)

The focus here is on workplace dilemmas and on the causes and effects of deliberate miscommunication.

To illustrate an ethical dilemma, you might use a familiar scenario like this one (based on Ed Henry's "Best of the Minivans." *Changing Times*, July 1990: 41–45):

> Assume you are Communications Director of a major automobile manufacturer in 1990. Nationwide, the market for minivans has been growing steadily for several years, and your company is just about to unveil its own minivan model, the Sidewinder, a "Luxury, All-Purpose, Family Van."

> Your department has been asked to write a complete product description of the Sidewinder. Versions of your descriptions will be used in various advertising literature, commercials, and in the annual report to stockholders. And so you are expected to persuasively present the "virtues" of this product.

> In preparation for writing, you review the research on minivans and speak with company engineers. Your investigation uncovers all kinds of facts, most notably these:

1. Minivans are big sellers among families with small children because they are easy to load and unload, can carry lots of kids and gear, and provide good gas mileage. In short, minivans are seen as a "practical choice" for families.

2. Among all motor vehicles, minivans have the fewest accident claims (presumably because they are usually driven by responsible adults transporting children).

3. Because they are classified by the government as "multipurpose vehicles" minivans are not required to meet the same safety standards required of passenger cars. And so, to hold down costs, minivans (including your company's Sidewinder) typically lack rear-seat shoulder belts, air bags, crash reinforcement in sides and roof, strong bumpers, rollover protection, head restraints, and other safety features. Consequently, minivan accidents, although statistically fewer than for other vehicles, can result in far more serious injuries.

Your employer's purpose in requesting this description is to "sell" the product.

Given the above facts, what should you do? Should you simply do as you are told? Do your specific responsibilities in this situation extend beyond completing your assigned task (i.e., promoting the product)? How much of the above material do consumers deserve to know? What should be emphasized? To whom do you owe the greatest loyalty here: yourself, the consumer, or your employer? What do you think will happen if you tell the whole truth? If you don't?

In a memo to your instructor give a detailed response to the above questions and comment on any other issues you can identify here. Be prepared to discuss your answers in class.

Besides illustrating a workplace dilemma, the previous scenario has elements that embody the dilemma we face as writing teachers who could presume to teach ethics at all: should we advance the organizational perspective (which tends to stress professional competence and the organization's welfare) or the academic perspective (which tends to stress social good)? One researcher points out that the first perspective engenders ethical equivocation, while the second imposes rarefied standards that are seen as unrealistic in the world of work. [See Gregory Clark's lucid and insightful article, "Ethics in Communication: A Rhetorical Perspective," *IEEE Transactions in Professional Communication* 30.3 (September 1987): 190–196.]

This chapter aims at a balance (albeit tenuous) between equivocal and polemical viewpoints by taking a descriptive, rather than prescriptive, approach to the ethics problem: namely, by examining the issues and inviting readers to draw their own conclusions.

Master Sheet 33 can be duplicated and handed out for any assignment.

Discussion of the Collaborative Project

The Ethics Collaborative Project (Master Sheets 34–38) produces much debate, labor, and energetic discussion. I consider it one of my most enjoyable course activities in recent memory—and students agree! We did find, however, the need to repeatedly orient our focus toward the ethical problems here.

In early drafts of their documents, teams tended to focus only on solving Killawatt's problems (with productivity, security, and so on), by merely tinkering with the original proposal; and so instead of lucid critical assessment of the actual and potential ethical violations, many of the initial responses seemed more an expression of company groupthink. But with some prodding during our workshops the class rapidly identified a whole range of ethical considerations in this scenario. Master Sheets 37 and 38 can be used at some point in the discussion (ideally at a later point) to summarize some of the major ethical concerns.

After two class periods, we ran out of time, but we all agreed that discussion here could have continued indefinitely.

Master Sheet 33

Checklist for Ethical Communication*

☐ Do I avoid exaggeration, understatement, sugarcoating, or any distortion or omission that leaves readers at a disadvantage?

☐ Do I make a clear distinction between "certainty" and "probability"?

☐ Am I being honest and fair?

☐ Have I explored all sides of the issue and all possible alternatives?

☐ Are my information sources valid, reliable, and unbiased?

☐ Do I actually believe what I'm saying, instead of being a mouthpiece for groupthink or advancing some hidden agenda?

☐ Would I still advocate this position if I were held publicly accountable for it?

☐ Do I provide enough information and interpretation for the audience to understand the true facts as I know them?

☐ Am I reasonably sure this document harms no innocent persons or damages their reputations?

☐ Am I respecting all legitimate rights to privacy and confidentiality?

☐ Do I inform the audience of the consequences or risks (as best I am able to predict) of what I am advocating?

☐ Do I state the case clearly, instead of hiding behind jargon and generalities?

☐ Do I give candid feedback or criticism, if it is warranted?

☐ Am I providing copies of this document to everyone who has the right to know about it?

☐ Do I credit all contributors and sources of ideas and information?

---

*This list largely was adapted from Judi Brownell and Michael Fitzgerald, "Teaching Ethics in Business Communication: The Effective/Ethical Balancing Scale," *The Bulletin for the Association of Business Communication* 55.3 (1992): 15–18; John Bryan, "Down the Slippery Slope: Ethics and the Technical Writer as Marketer," *Technical Communication Quarterly* 1.1 (1992): 73–88; Richard L. Johannesen, *Ethics in Human Communication*, 2nd Edition. (Prospect Heights, Illinois: Waveland, 1983: 21–22); Charles U. Larson, *Persuasion: Perception and Responsibility*, 7th Edition. (Belmont, CA: Wadsworth, 1995: 39); Stephen H. Unger, *Controlling Technology: Ethics and the Responsible Engineer* (New York: Holt, 1982: 39–46); George Yoos, "A Revision of the Concept of Ethical Appeal," *Philosophy and Rhetoric* 12 (Winter 1979): 41–58.

Master Sheet 34

Ethics Collaborative Project

Workplace decisions with ethical implications often are made *collaboratively* by members of a project or management team. This project offers practice in collaborative decision making on a growing ethical problem spawned by new technology.

Divide into groups. Assume you are a junior manager at Killawatt, a major producer of consumer electronics, from digital watches to cordless phones to computer games to the newest gadgetry. Your team is composed of managers from various departments such as research and development, computer services, human resources, accounting, production, and corporate relations. You have been brought together to tackle the following problem.[1]

Analyze the Problem

Killawatt is facing a threefold management problem:

1. A two-million dollar loss in the recent quarter, caused by an economic down-turn, necessitates severe austerity measures. Upper management wants to eliminate internal losses from such sources as:

   - excessive and unauthorized use of photocopy and fax machines, digitizers, laser printers, Internet browsers, and other equipment
   - personal phone calls on the WATS line
   - personal or excessive use of database retrieval services such as CompuServe and Dialog (costing as much as $5.00 per minute), for stock market quotations, investment information, and the like
   - extra-long lunch hours and coffee breaks
   - personal use of company cars and trucks
   - unlimited purchases of company products at a 40 percent employee discount

In rosier economic periods, all these "perks" were tolerated by Killawatt. But today's hard times make such benefits unaffordable.

2. A more crucial and long-term problem for the company is to find ways of identifying its most and least productive employees. Presumably, this step could not only increase productivity, but might also provide a more selective basis for salary increases, bonuses, promotions, and even any layoffs that might later be unavoidable.

---

[1]Adapted from "Monitoring on the Job," by Gary T. Marx and Sanford Sherizen (Nov./Dec 1986): 63–72. Reprinted with permission from *Technology Review,* © 1986.

Master Sheet 35

Ethics Collaborative Project

3. A third and somewhat related problem is for Killawatt to plug all security leaks. In the past two years, major company breakthroughs in product technology have been leaked to competitors. And details of the company's recent fiscal problems have been leaked to the press, hurting stock value and scaring investors.

In light of these problems with perks, productivity, and security, Killawatt is exploring ways to monitor its employees. (Monitoring equipment, although expensive, could be capitalized, depreciated, or written off as a business expense; resulting tax benefits would largely offset equipment costs.) At this preliminary stage, you have been asked to assess the feasibility of such a plan in terms of its impact on employee relations.

Like many co-workers, you have heard your share of horror stories about perfectly "legal" monitoring in some other companies. For example, security officers for one government contractor rummage through employees' desk drawers after work hours. Abuses at other companies include:

- Random checking of employees' computers for unauthorized files (such as personal correspondence, coursework, home designs, and the like).

- Devices that monitor computer workstations by keeping track of the number of keystrokes, errors, corrections, words typed per minute, customers processed per hour—even the number and length of visits to the bathroom.

- Companywide and random polygraph (lie detector) testing to identify dishonest employees.

- Cameras or microphones to keep track of productivity, behavior, and even conversations.

- Telephone devices that keep track of each call: the extension from which it was made, the number dialed, time and length of call, and even a recording of the conversation. Such devices presumably help expose security leaks by keeping track of who is talking to whom about what.

- Programs that tell workers how their performance is measuring up against co-workers', or some with "relaxation" sounds or other subliminal messages designed to increase concentration or speed or morale.

While advocates of monitoring at Killawatt acknowledge the potential for abuse, they argue that a conscientious program could benefit the employees as well as the company. Besides keeping everyone "honest," monitoring could provide an objective measure of productivity, without the interpersonal and sometimes discriminating clashes between supervisor and employee. The technology, in effect, would ensure equal treatment across the board. Annual performance appraisals could then be based on reliable criteria that would enhance a supervisor's "subjective" impressions of the employee's contribution.

Master Sheet 36

Ethics Collaborative Project

But these arguments have not persuaded you or your team members. In your view the plan raises troubling ethical questions:

- What are the employees' rights in this issue?
- Is the plan fair?
- Could the monitoring plan backfire? Why?
- How would the plan affect employees' perceptions of the company?
- How would the plan affect employee morale and loyalty and productivity?
- What should be done to avoid alienating employees?
- Are there acceptable alternatives to monitoring?

Your team struggles with these and other questions you identify when you meet to plan your response. How will your team respond? What position will you advocate? What are the human costs and benefits of the proposed plan? What obligations, ideals, and consequences are involved? What plan can your team offer instead? How can you be persuasive here?

Plan Your Response
As your team plans its response to management, think of key points you want to emphasize.

Analyze Your Audience(s)
Your primary audience consists of executives and managers, many who see nothing wrong with monitoring. If your team's memo (followed by a meeting) manages to persuade your bosses, then you will be expected to win employee acceptance of your plan—and so your secondary audience will consist of employees who mostly are strongly opposed to any type of monitoring. Prepare audience-and-use profiles (pages 28–30) of each audience.

Prepare and Present Your Documents
In a memo to your bosses, argue against their plan and make a *persuasive* case for your alternative. Be explicit about your objections and about the *ethical* problems you envision. Brainstorm for worthwhile content, do any research that may be needed, write a workable draft, and revise using the Ethics Checklist (page 78).

Assume that your argument succeeds with your bosses, and prepare a memo to employees that is reassuring and persuasive in eliciting their cooperation with your plan.

Appoint a team member to present the finished documents (along with complete audience profiles) for class evaluation and response.

Master Sheet 37

Partial Response for Ethics Collaborative Project

Some Objections to Employee Monitoring

These arguments from advocates of monitoring haven't persuaded you or your team members. In your view, this "omniscient" technology would violate employees' privacy, would alienate them, and would undermine their morale and loyalty to the company, ultimately hurting rather than helping productivity. (It's hard to feel like a trusted team player in an impersonal work environment riddled with suspicion.) In fact, you're convinced that the whole "piecework" climate of monitoring would emphasize quantity at the expense of quality, and might even encourage sabotage or further security leaks. Moreover, some of Killawatt's most loyal, talented, and sought-after employees could be driven to other companies.

As you plan your response to management, you think of some key points you want to emphasize. (See Master Sheet 38.)

Master Sheet 38

Partial Response for Ethics Collaborative Project
(continued)

Key Points of the Objection and Counterproposals

- Employees should have a direct voice in setting any guidelines.

- They must be fully informed of the problem and the proposed policy.

- Individual privacy must be respected and preserved at all costs.

- If some type of monitoring turns out to be unavoidable,

  (a) the program should have employee consensus;

  (b) employees should know exactly when they are being monitored;

  (c) employees should have access to any personal information gathered concerning them;

  (d) otherwise, the information should be strictly confidential;

  (e) any monitoring should be evenly applied to *all* levels of employees.

- Killawatt might avoid monitoring altogether by

  (a) asking for *voluntary* compliance with cost-cutting and security measures;

  (b) using coded access cards for photocopy machines;

  (c) restricting direct dialing to local phone calls, with long distance only available through the switchboard;

  (d) asking employees to shorten breaks and to stop taking advantage of other perks, as part of a "companywide conservation and recycling drive" for cutting costs on electricity, supplies, and so on;

  (e) developing an incentive program for cost-cutting suggestions from employees;

  (f) putting up posters at employee entrances with such messages as "Don't be late; others are depending on you;"

  (g) scheduling meetings and tasks immediately following lunch hour (no one likes to walk into a full conference room or a meeting in progress).

- To prevent security leaks, ask employees to sign a security pledge, or provide a hotline to enable employees to report offenders.

- To decrease inconsistent productivity among employees, institute a "self-monitoring" system, say, with weekly reports using prepared forms.

Master Sheet 39

Chapter 5 Quiz

Name_____Section _____

Indicate whether statements 1–6 are TRUE or FALSE by writing *T* or *F* in the blank.

1. _____ To be effective, a message must be ethical.

2. _____ The law offers full protection for employees who take an ethical stand against an employer.

3. _____ The *Challenger* disaster was an unavoidable consequence of mechanical failure.

4. _____ In ethical organizations, "groupthink" is preferable to "team play."

5. _____ In reports to their superiors, subordinates tend to suppress bad news.

6. _____ The law provides adequate guidelines for ethical behavior.

Choose the letter of the expression that best completes each statement.

7. _____ Lies that are "legal" in the workplace include all of the following, except (a) promises you know you can't keep, (b) assurances you haven't verified, (c) credentials you don't have, (d) broken contractual promises, or (e) inflated claims about your commitment.

8. _____ Ethical employees always owe their greatest loyalty to (a) themselves, (b) clients and customers, (c) their company, (d) co-workers, or (e) none of these.

9. _____ Ethical employees stand a better chance of speaking out and surviving if they avoid everything except (a) overreacting, (b) procrastination, (c) keeping a "paper trail," (d) overstating the problem, or (e) crusading.

10. _____ Among managers polled nationwide, those who felt pressured by their company to compromise their ethical standards numbered (a) fewer than 10 percent, (b) greater than 50 percent, (c) over 90 percent, (d) 25 percent, or (e) 40 percent.

CHAPTER

6

Solving the Collaboration Problem

The eighth edition of *Technical Communication* includes collaborative projects at the end of most chapters. Additional projects can be found in this manual (See Table of Master Sheets, p. v). These assignments serve a number of purposes:

1. They offer practice in collaborative writing, a frequent arrangement in the workplace, especially for long and complex documents. Team members face many challenges, including assigning and completing tasks; making timetables; meeting deadlines; organizing data from various sources; achieving a uniform and appropriate style in a team-written document; evaluating and being evaluated by peers; learning to manage a project—and to be managed. In short, students learn the meaning of working together toward a goal.

2. Collaborative projects confront students with interpersonal problems that almost invariably crop up during any demanding group effort: achieving consensus and cooperation; overcoming personality differences; dealing with poorly motivated or domineering colleagues; achieving fair distribution of labor, and so on. In short, students learn how to get along to get the job done.

3. Collaborative projects enable weaker writers to benefit by working with stronger writers in planning, researching, drafting, and revising a document.

4. Collaborative projects often enable team members to accomplish a broader range of tasks than any student could accomplish individually.

Admittedly, any collaborative project can have disappointing results for instructor and students alike. But even a project that proves disastrous can go a long way toward teaching students something about accountability and shared responsibility.

The success of a collaborative project depends primarily, I think, on the group's motivation. One way to stimulate motivation is by asking each group to elect a project manager. The manager will then assign tasks and, at project's end, will evaluate—in writing each member's contribution. Each member, in turn, will evaluate the manager.

In Chapter 6, students will find useful guidelines for writing collaboratively. Included are a management plan sheet and a form for student evaluation of team members (Master Sheets 40 and 41).

Master Sheet 40

Plan Sheet for Managing a Collaborative Project

Management Plan Sheet

Project title:
Audience:
Project manager:
Team members:
Purpose of the project:

. . . .

Specific assignments	**Due dates**
Research:	Research due:
Planning:	Plan and outline due:
Drafting:	First draft due:
Revising:	Reviews due:
Preparing final document:	Revision due:
Presenting oral briefing:	Final document due:
	Progress report(s) due:

. . . .

Work Schedule

Group meetings:	Date	Place	Time	Note-taker
#1				
#2				
#3				
etc.				

Meetings with instructor:
#1
#2
etc.

. . . .

Miscellaneous:

How will disputes and grievances be resolved?

How will contributions be evaluated?

Other matters (Internet searches, email routing, computer conferences)?

Master Sheet 41

Sample Form For Evaluating Team Members

Performance Appraisal for _____

(Rate each element as *[superior]*, *[acceptable]*, or *[unacceptable]* and use the "Comment" section to explain each rating briefly)

- *Cooperation:* [_____]
    Comment:

- *Dependability:* [_____]
    Comment:

- *Effort:* [_____]
    Comment:

- *Quality of work produced:* [_____]
    Comment:

- *Ability to meet deadlines:* [_____]
    Comment:

_____
Project Manager's Signature

Master Sheet 42

Chapter 6 Quiz

Name_____ Section _____

Indicate whether Statements 1–7 are TRUE or FALSE by writing *T* or *F* in the blank.

1. _____ In collaborating to produce a document, all members of a collaborative team participate in the actual "writing."

2. _____ Conflict in a collaborative group can be productive.

3. _____ A collaborative group functions best when each of its members has equal authority.

4. _____ Workplace surveys show that people view meetings as a big waste of time.

5. _____ Conflict in collaborative groups can increase when the group transacts exclusively online.

6. _____ "Reviewing" is a more precise term for "editing."

7. _____ In mixed-gender groups, assertive females are considered highly persuasive.

In 8–10, choose the letter of the expression that best completes the statement.

8. _____ Sources of conflict in collaborative groups include (a) interpersonal differences, (b) gender differences, (c) cultural differences, (d) all of these, or (e) b and c.

9. _____ Strategies for creative thinking include all of the following except (a) brainstorming, (b) brainwriting (c) brain scanning, (d) mind mapping, and (e) storyboarding.

10. _____ Electronically mediated collaboration is preferable when (a) people don't know each other, (b) the issue is sensitive or controversial, (c) it is important to neutralize personality clashes, (d) all of these, or (e) none of these.

PART

II

Retrieving, Analyzing, and Synthesizing Information

This section treats research as a deliberate inquiry process. Students learn to formulate significant research questions; to explore, interpret, and document their findings; and to summarize for economy, accuracy, and emphasis.

Thinking Critically about
the Research Process

Chapter 7 presents an overview of critical thinking in designing a legitimate inquiry: asking the right questions, focusing on essential sources, and evaluating and interpreting findings. This chapter (along with Chapters 8, 9, and 10) can serve as the basis for the semester's major writing assignment. It may be incorporated into the approach to the formal analytical report discussed in Chapter 25 or the proposal discussed in Chapter 24.

"The Purpose of Research" on Master Sheets 43 and 44 is especially important to lower-level students, who too often equate research with the old high-school paper about life on Mars or the Bermuda Triangle. Students need to understand that research is done for a purpose, that it is not just an academic exercise, and that the information uncovered will be put to practical use.

You might give students a scenario like this: "You're thinking of starting an investment group with friends in your community and you need to decide whether to invest in mutual funds, municipal bonds, real estate, long-term savings certificates, precious metals, or the like. Trace the past five years in each of these investment areas, comparing overall return on every dollar invested." Better yet, ask each student to think up a research topic that would yield information of practical use—as illustrated in Master Sheets 45 and 46.

For any research project, students should follow a well-defined schedule for completing the various tasks—as outlined in exercise 1. Planning helps avoid the last-minute all-nighter, and the poor writing that inevitably results.

Master Sheets 43–44

The Purpose of Research

The purpose behind all research is to arrive at an informed opinion, to establish the conclusion that has the greatest chance of being valid.

We might have uninformed opinions about political candidates, kinds of cars, controversial subjects like abortion and capital punishment, or anything else. Opinions are beliefs that are not proven but seem to us to be true or valid. Without a basis in fact, opinions are uninformed, disputable, and subject to change in the light of new experience. Sometimes we forget that many of our opinions rest on no objective data. Instead, they are based on a chaotic collection of the beliefs reiterated around us, notions we've absorbed from advertising, things we've read but never checked for validity. Commercials especially are designed to manipulate the consumer's uninformed opinion.

Any claim is valid only insofar as it is supported by facts, what we know from observation or study. An opinion based on fact is more legitimate than an uninformed opinion. We must often consider a variety of facts. Consider a commercial claim that Brand X toothpaste makes teeth whiter. Although this claim may be true, a related fact may be that Brand X toothpaste contains tiny particles of ground glass, thereby harming more than helping teeth. The second fact may change your opinion. Similarly, a United States senator may claim an environmental commitment, but may have voted against important environmental legislation. If you want to know whether to believe the senator's claim, you need to establish the facts. You want your opinion to rest on verifiable information, and not merely on the claim.

Facts also affect the quality of personal decisions. Before studying for a career, investigate job openings, salary range, and requirements; better yet, interview someone who has such a job. Conduct the same kind of research before buying a new car or emigrating to Australia. Although the facts you discover may contradict your original opinion, they will enable you to make an informed decision.

On the job, only the rare important decision is made without some research. And the results of any research are almost inevitably recorded in a report.

Master Sheet 44

Research is a prerequisite to starting your own business. Before locating and opening a motorcycle shop, you should investigate the local marketing possibilities and consumer profiles. In what age range and socioeconomic level are most motorcycle buyers? What percentage of your local population are such consumers? Are motorcycle sales increasing or decreasing nationwide? How many motorcycles are now registered in your town? Are any similar shops in the area?

If answers to these and other questions are encouraging, you now must choose the best location for your business. First, identify the type and amount of space needed. Next, inspect all available local rentals, comparing rental costs and the advantages of each. Now, choose the best location, basing your choice on rental cost, customer traffic, parking area, and types of neighboring businesses. Should you locate downtown on Main Street, in the shopping mall just outside the center of town, or in a small, freestanding building two miles out of town, where motorcycle noise will bother no one?

These are just a few of the questions that anyone starting such a business must answer accurately. And your answers are found in the facts you uncover in library materials, personal observation of possible sites, interviews and meetings with small-business advisory services, questionnaires distributed to a cross-section of the consumer public, letters of inquiry to motorcycle manufacturers, and a review of records (registry records, census records, sales records). Finally, you will present your purpose, findings, and conclusions in a report to the local bank to justify your request for a business loan.

If you join an organization instead, you may be asked to investigate and report on the feasibility of marketing a new product, the benefits of a merger with another corporation, the value of a proposed building site for a new branch, the advisability of purchasing a new piece of medical or dental equipment, and so on. Anyone in a responsible position—store owner, technician, or executive—needs to keep up with new developments. Research is by no means an exercise for eccentric scientists or bleary-eyed scholars.

Master Sheet 45

Examples of Research Topics

Here are some possible research topics:

a. Survey student, faculty, and administration about some proposed curriculum change or about some other controversial campus issue. Compare your findings with nationwide or statewide statistics about attitudes on this issue.

b. As much as 30 percent of groundwater in some states is contaminated. Find out how the quality of local groundwater measures up to national averages. Has the quality increased or decreased over the last ten years? What are the major elements affecting local water quality? What is the outlook for the next decade? Can local residents feel safe drinking tap water?

c. Identify the main qualities employers seek in job applicants. Have employers' expectations changed over the last ten years? If so, why? What is the chance that your generation will face six or seven career changes in your lifetime? What should people do to prepare?

d. Find out which geographic area of the United States is enjoying the greatest prosperity and population growth (or which area is suffering the greatest hardship and population decrease). What are the major reasons? Trace the recent history of this change.

e. Older homes can present a frightening array of toxic hazards: termite spray, wood preservatives, urea formaldehyde insulation, radon gas, chemical contamination of well water, lead paint, asbestos, and so on. Research the major effects of these hazards for someone who is thinking of buying an older home.

f. How safe is your school (or your dorm) from "sick-building syndrome"? Are there any dangers from insulation, asbestos, art supplies, cleaning fluids and solvents, water pipes, or the like? Find out, and prepare a report for your classmates.

g. Which area of your state has the cleanest air and groundwater, and which has the most polluted? Write for someone looking for the safest place to raise a family.

h. Has acid rain caused any damage in your area? Write for classmates.

i. Can peanut butter, black pepper, potatoes, or toasted bread cause cancer? Which of the most common "pure" foods can be carcinogenic? Find out, and write a report for the school dietitian.

j. Are there any recent inventions that could help decrease our reliance on fossil fuels in ways that are economically feasible and practical? Find out, and prepare a report for your U.S. senator.

Master Sheet 46

Examples of Research Topics (continued)

k. What is the very latest that scientists are saying about the implications of global warming from rain forest destruction and ozone depletion? Find out, and prepare a report to be published in a national magazine.

l. Computer screens: How dangerous are they? What does the latest research indicate? Your company wants to know if it can take any precautions to avoid risks to employees and future lawsuits.

m. Say you work for a "Think-Tank" researching the following issues. Or you work for a U.S. senator who wants to introduce legislation to curb the abuses. What kinds of privacy violations does present law allow in the workplace? What legislation is pending?

n. Does alcohol consumption have any effect on academic performance? What do the latest studies indicate? Find out, and prepare a report for publication in your campus newspaper.

o. Assume you are preparing for a job interview with an organization hiring in your major. Learn all you can about the organization's history, management style, growth prospects, products and services, mergers, multinational affiliations, ethical record, environmental record, employee relations, and other important features. Prepare a report that will be available for other students in your major who might have an interest in this organization.

Master Sheet 47

Chapter 7 Quiz

Name _____ Section _____

Indicate whether statements 1–7 are TRUE or FALSE by writing *T* or *F* in the blank.

1. _____ Effective research eliminates contradictory conclusions.

2. _____ Expert testimony usually offers a reliable "final word."

3. _____ Web pages usually are dependable sources for information that offers both depth and quality.

4. _____ Decisions in the research process are recursive.

5. _____ The research process can be visualized as being a "looping" structure.

6. _____ In research, "balance" and "accuracy" are synonymous.

7. _____ An ethical researcher reports all points of view as if they were equal.

In 6–10, choose the letter of the expression that best completes each statement.

8. _____ At its deepest level, secondary research examines (a) trade and business publications, (b) the popular press, (c) tabloids, (d) specialized and government sources, or (e) electronic newsgroups.

9. _____ Effective research depends on all of the following except (a) finding a definite answer, (b) sampling a full range of opinions, (c) getting at the facts, (d) achieving sufficient depth, or (e) asking the right questions.

10. _____ Likely sources for relatively impartial views are (a) trade and business sources, (b) Web pages, (c) specialized and government sources, (d) all of these, or (e) none of these.

CHAPTER
8

Exploring Hardcopy,
Internet, and Online Sources

Freshmen and sophomores benefit from a library tour. Juniors and seniors already should have a working knowledge of the library, but likely need some orientation to electronic information services. Supplement the library visit with hands-on experience by assigning selected exercises.

To create a context for and to supplement the sample topics offered in the Collaborative Projects (or for those you assign or students themselves propose), you might ask students to assume they have been hired by a company or a college to search for data that will be used by others in making a decision. (Perhaps a company vice-president wants to know why some companies hesitate to automate their offices, or why many managers resist the notion of automated offices. Or perhaps an executive in a small chemical company wants to know how innovative companies are solving the problem of toxic-waste disposal. The possibilities for realistic scenarios are endless.) You might mention that even the best trial lawyers spend a good deal of time in the library doing homework before presenting data upon which the jury will base its verdict.

Emphasize the importance of knowing where and how to find the information one needs *when* one needs it.

Master Sheet 48

Chapter 8 Quiz

Name _____ Section _____

Indicate whether statements 1–8 are TRUE or FALSE by writing *T* or *F* in the blank.

1. _____ Reference books are a good place to begin library research.

2. _____ An electronic search of the literature usually eliminates the need for searching hardcopy sources.

3. _____ In general, the more accessible the source, the less valuable it is likely to be.

4. _____ Only published works receive copyright protection.

5. _____ Material obtained from the Internet is protected by copyright.

6. _____ An intranet is an in-house network that uses Internet technology.

7. _____ For distributing information, "push" strategies are more efficient than "pull" strategies.

8. _____ Material in the "public domain" is not protected by copyright.

In 6–10, choose the letter of the expression that best completes each statement.

9. _____ Computerized databases rarely contain entries published before (a) the mid-1960s, (b) the 1920s, (c) the late 1970s, (d) 1990, or (e) 1941.

10. _____ The major access tool for government publications is (a) *The Monthly Catalog of the United States Government,* (b) *Vital Federal Documents,* (c) *Selected Government Publications,* (d) Congressional Record, or (e) Library of Congress.

CHAPTER
9

Exploring Primary Sources

Students at any level should be encouraged (or required) to compose questionnaires or plan interviews as part of their data bank for long reports. Your advice about the rough drafts of interview questions or sample questionnaires will be a big help. This is a good occasion to have individual conferences.

Master Sheet 49

Chapter 9 Quiz

Name _____ Section _____

Indicate whether statements 1–10 are TRUE or FALSE by writing *T* or *F* in the blank.

1. _____ To measure exactly where people stand on an issue, use closed-ended survey questions.

2. _____ Instead of writing out your interview questions, create a relaxed atmosphere by memorizing the questions beforehand.

3. _____ Use closed-ended survey questions to eliminate biased responses.

4. _____ Get the most difficult, complex, or sensitive questions out of the way at the beginning of the interview.

5. _____ Take plenty of notes during the interview.

6. _____ Begin a survey with the easiest questions.

7. _____ Direct observation is the surest way to eliminate bias in research.

8. _____ To eliminate the potential for error, have your survey designed by a professional.

9. _____ Generally, the most productive way to conduct an interview is by phone.

10. _____ A "sample group" should never be randomly chosen.

Evaluating and
Interpreting Information

This chapter continues the emphasis on critical thinking: in evaluating sources and evidence and in interpreting the material.

Students need reminding that (1) we do research not just to find answers but to find answers that stand the best chance of being correct; and (2) as *consumers*, as well as producers, of research, we need to be able to evaluate and interpret the findings accurately.

To work with their information from a critical perspective, students should be able to evaluate the material for validity, reliability, and certainty; to recognize the influence of bias (for example: in database sources, direct observation, or interpretation); to avoid causal and statistical fallacies; and to reassess their research methods and reasoning before settling on a conclusion.

Suggested responses to exercise 4.

4.a. One alternative explanation is that people are living longer and thus are more likely to die of cancer and that cancer today rarely is misdiagnosed—or mislabeled because of stigma.

4.b. Perhaps the most difficult and hopeless cases are sent here from other hospitals.

4.c. Certain cultures are reluctant to acknowledge emotional illness. Moreover, countries with low rates of separation and divorce also have low rates of depression.

4.d. Sklaroff and Ash point out that "this doesn't say much about the effect of pot on brain power, but it does describe a small subset of teenagers who grew up in environments where, directly or not, both academic achievement and marijuana use were condoned."

4.e. This item is self-explanatory!

Suggested response to collaborative project 3.
An accurate estimation would have to compare teachers' wages with the average wage of *other workers who have similar minimum qualifications:* namely, a four-year college degree (often, graduate degrees as well), internship experience, and certification in their specialty.

Master Sheet 50

Chapter 10 Quiz

Name _____ Section _____

Indicate whether statements 1–7 are TRUE or FALSE by writing *T* or *F* in the blank.

1. _____ The most recent information is almost always the most reliable.

2. _____ Research is most effective when it achieves "certainty."

3. _____ In statistics, the *mode* is the value that occurs most often.

4. _____ Numbers tend to be less misleading than words.

5. _____ Personal bias among researchers is inescapable.

6. _____ "Framing" offers an ethical way to present the facts.

7. _____ "Correlation" is a more precise term for "causation."

In 8–10, choose the letter of the expression that best completes each statement.

8. _____ The basic criteria by which we measure the dependability of any research are (a) timeliness and efficiency, (b) validity and reliability, (c) conciseness and emphasis, (d) relevance and focus, or (e) all of these.

9. _____ Valid research typically produces (a) a conclusive answer, (b) a probable answer, (c) an inconclusive answer, (d) a or b, or (e) a, b, or c.

10. _____ Indicators of quality for a Web site include all of the following except (a) links to reputable sites, (b) material that has been peer reviewed, (c) options for contacting the author or organization, (d) presentations that include graphics, video, and sound, or (e) objective coverage.

11

Summarizing Information

Summaries are a logical early assignment because they (1) apply the principles of audience analysis and clear prose writing in specific practices; (2) provide a means for writing efficiently and concisely; (3) improve note-taking and study skills in other courses by helping students recognize and differentiate major from minor points; (4) provide practice in writing informative paragraphs that are unified and coherent; and (5) teach students to extract the essential message from a longer piece and to communicate it intact, helping to develop a skill that will be used throughout the semester.

You might supplement the editing advice on page 176 by referring to corresponding sections of Chapter 13 and the Appendix.

Spend time discussing the summary of "U.S. Nuclear Power Industry" and its various versions before having students write their own summaries. For the final revision you might wish to review the section on transitions in Appendix C. Explain that even a summary can be compressed, depending on one's audience and purpose.

Explaining the difference between summaries and abstracts is a sticky problem, primarily because of the overlapping nomenclature:

> summary   =   informative abstract
> abstract   =   descriptive abstract
> summary   =   abstract

Perhaps the most sensible approach to this problem is to think of an abstract as discussing what the original is about and a summary as discussing both what the original is about and the major facts it covers.

Tell students that summaries of any kind in the workplace usually are written only for long documents, not for letters or memos.

Exercise 1 works well and quickly in class. Exercise 5 should be done at home and brought in for small-group workshops using revision checklists. After allowing ample time for the workshop, ask the class to nominate a superior piece, then show it on the opaque projector for further criticism. Ask students to revise their summaries at home before you see them—according to their editors' suggestions. If your class meets once weekly, you might use exercise 5 as an in-class exercise. Have students prepare at home by reading, highlighting, and listing major points. Then have small groups prepare a composite summary in class.

As an additional assignment, ask students to summarize a short chapter from one of their textbooks and to turn it in with a photocopy of the original. Suggested responses to exercise 5 are on Master Sheets 51–52. Notice how the original text has been somewhat restructured in this informative abstract, for greater access and sharper emphasis—but without distorting the original message. Notice also the clarifying definition of "contraindications" from paragraph 16 in the original and "cardiovascular disease" from paragraph 1. Be sure to point out that the contents of this abstract are designed specifically for the needs of a general audience (as described in the exercise 5 scenario). For some other audience (say, student nurses or medical personnel), the choice of details and emphasis presumably would be different.

Master Sheet 51

Response To Exercise 5, Chapter 11

Informative Abstract (roughly 23 percent of original length)
Despite more than a century of routine use, and some 80 billion tablets consumed yearly in the U.S., aspirin continues to be tested as a new treatment for conditions including cardiovascular (heart and blood vessels) disease, cancer, migraine headache, and high blood pressure in pregnancy.

In addition to easing aches and pain and reducing fever and arthritic inflammation, aspirin interferes with blood clotting. By dissolving blood clots in the arteries that feed the heart muscle, aspirin helps prevent the heart attacks that strike 1,250,000 victims yearly in the U.S. and kill some 500,000.

Studies show that aspirin substantially reduces the risk of second heart attacks and of first attacks among people who experience chest pain (angina pectoris) beforehand. But while prolonged aspirin use seems to prevent first heart attacks it slightly increases stroke risk and causes no apparent reduction in overall deaths from cardiovascular disease.

Doctors recommend low-dose aspirin therapy for men aged 40 and over who are at risk for heart attack and who display no other conditions that could be aggravated by aspirin, such as clotting disorders and stomach ulcers.

Other uses being studied:
- treating migraine headache
- fighting periodontal disease by improving circulation to gums
- preventing some types of cataracts
- lowering risk of recurring colon and rectal cancer
- controlling high blood pressure (preeclampsia) during pregnancy

Master Sheet 52

Response To Exercise 5, Chapter 11 (continued)

But none of these uses is yet proven safe or effective. And some people are likely to misuse aspirin because of mistaken assumptions about its effectiveness.

Self-treatment—even at low doses—can cause serious side effects:

- allergic reactions
- nausea, heartburn, and stomach pain
- worsening of high blood pressure, liver or kidney disease, or peptic ulcer
- hemorrhagic strokes (bleeding into the brain) or other internal bleeding (say, stomach or intestine), caused by aspirin's anti-clotting properties
- Reyes syndrome, a rare but sometimes fatal affliction in children and teenagers who have flu or chicken pox

Despite its broad potential, we should use aspirin cautiously until more evidence is in. In fact, the Food and Drug Administration suggests that we consult a physician before taking aspirin for any new or long-term uses.

Descriptive Abstract
The medical benefits of aspirin are identified and balanced against potential side effects.

Master Sheet 53

Chapter 11 Quiz

Name _____ Section _____

Indicate whether statements 1–6 are TRUE or FALSE by writing *T* or *F* in the blank.

1. _____ A summary never should be longer than 5 percent of the original.

2. _____ Whenever possible, you should insert your personal comments about material in the summary.

3. _____ Generally you should summarize someone else's material in your own words.

4. _____ New employees often summarize professional literature for decision makers.

5. _____ The length of your summary should be your primary interest.

6. _____ More people are likely to read your summary than any other part of your report.

In 7–10, choose the letter of the expression that best completes each statement.

7. _____ Above all, a good summary is (a) brief, (b) grammatical, (c) concise, (d) entertaining, or (e) obtuse.

8. _____ For a summary, the essential message does not include (a) controlling ideas, (b) conclusions and recommendations, (c) examples and visuals, (d) major findings, or (e) key statistics.

9. _____ To avoid confusing anyone in a large and unspecified audience, make your summary (a) highly technical, (b) nontechnical, (c) as brief as possible, (d) as entertaining as possible, or (e) graphically engaging.

10. _____ A descriptive abstract presents (a) the most general view, (b) the most detailed view, (c) the essential message from the original, (d) a visual depiction, or (e) the pedantic perspective.

PART

III

Structural
and Style Elements

This section shows students how to organize and express a message to meet readers' specific needs and expectations. Students learn to control their material and to develop a style that connects with readers.

12

Organizing for Users

The intention here is to show that a well-organized report doesn't just *happen*—it evolves from a careful plan.

Since partition and classification in isolation are abstract concepts, you might emphasize the dividing we do every day, asking students to come up with examples for a list you make on the board. Or ask them to devise a hypothetical situation in which classification and partition are important, as in the following supermarket example:

> If you are designing a new supermarket, for example, you must first partition the whole market into parts: display and shopping area, receiving and storage area, meat refrigeration and preparation area, and so on. Next, you need to sort your inventory by grouping the thousands of items into smaller classes according to their similarities: frozen foods, dairy products, meat, fish, and poultry, and so on.
>
> In turn, you partition each of these sections, say dividing "meat" into three smaller groups: beef, pork, and lamb. Under these headings, you will group the cuts of meat (steaks, ribs, and so on) in each category. You might carry the division further for some meat products, such as types of ground beef: regular, lean, diet lean.
>
> This kind of dividing and grouping continues until you have enough categories or classes to sort the hundreds of crates and cartons of inventory that sit in your receiving and storage area. You have used division and classification to divide your store into parts and to sort your inventory into classes.

Explain that because most people have trouble organizing their writing, these dividing strategies are intended to sharpen skills that will be used later in planning, outlining, and writing longer reports, and in sorting out research data.

For basic students, Master Sheets 54 and 55 show the practical uses of partition and classification.

The importance of outlining for writing students cannot be overemphasized. Both upper- and lower-level students—as well as most working professionals I've encountered—have the same need for intensive practice. As students work through their own outlines and criticize those of others, they begin to see writing as a *process*—a procedure that is deliberate and fully planned—instead of an exercise in which they empty their heads onto the page or blindly follow a pencil from line to line. Year after year, my students have rated learning to outline among their most valuable experiences in the course.

The Collaborative Projects are effective in class for hands-on practice in the applied thinking that produces a good outline. For Project 1, you might ask all groups to work on the same topic chosen from the list. This arrangement leads to a lively discussion session of samples placed on the board.

You probably will want to include on-the-board outlining workshops with later assignments in description, process explanation, research, and analysis. As an optional exercise, ask students to bring in an outline for a paper over which they've been struggling for another course and to work it out, with the class, on the board.

Stress that outlines come in all shapes and sizes, and, like building plans for a house, the outline can always be revised as needed. The neat and ordered outline shown in the text represents the *product* of outlining, not the *process*. Beneath any finished outline (or any finished document) lie pages of scribbling and things crossed out, jumbled lists, arrows, and fragments of ideas. Writing begins in disorder. Messiness is a natural and often essential part of writing in its early stages.

A visual extension of the outline is the storyboard, especially useful for cutting-and-pasting work in a collaborative group.

The students' early writing should be measured not by sheer volume, but rather by the level of deliberate and conscious decisions the students make. If students can learn to compose *one good paragraph,* then they've made real progress in grasping the principles of *all* good writing.

Recent theorists argue that a paragraph alone, in most cases, does not constitute a complete discourse. Nonetheless, a standard support paragraph is an organized and recognizable unit of meaning that, for the purpose of illustration, can be considered a discrete discourse. For that matter, even in a long discourse, meaning is made clear *within* paragraphs, where each main supporting idea is developed and clarified. Working with paragraphs as "units of meaning," students learn the importance of making their own meaning explicit instead of expecting readers to figure it out for themselves.

The bulk of problems with structure and content occur at the paragraph level, especially with paragraphs that are poorly developed or that are disorganized, arbitrary blocks of sentences on a page. Even writers with fluency and imagination often lack basic paragraph logic. The standard paragraph, then, is an appropriate beginning model for strong and weak writers alike. Even as a partial discourse, this model teaches essential features of rhetorical awareness: recognizable beginnings, middles, and endings; clear and distinct main points; convincing support; appropriate amounts of generality and abstraction; unity and coherence. These are features of *all* effective discourse, regardless of length.

Because they already have studied decisions about purpose, audience, and content, students can now appreciate how decisions about organization can help writers connect with their audience—how decisions about *what to say* are related to decisions on how to *organize.*

Much of the emphasis here is on topic sentences because this is precisely where writers must discover their exact meaning in order to compose a worthwhile and sensible paragraph. Derived from a solid topic sentence (and an adequately informed writer), other parts of the paragraph have a way of taking care of themselves. And, for the reader, the topic sentence serves as introduction and orientation to a unit of meaning. Here—and throughout the course—I emphasize that any introduction (to a paragraph, a letter, or a document of any length) should focus on what the reader needs first.

For lower-level students, you might compose a paragraph in class. Have them complete the whole process collectively, while you record it on the chalkboard—from audience analysis to brainstorming to writing and revising successive drafts.

Make time for editing workshops on the paragraphs students compose at home. These short and manageable units of discourse offer a vivid and concrete view of the role of decision making in drafting, editing, and revising various messages for various audiences.

Answers to Exercise 2

1. chronological
2. chronological
3. problem-causes-solution
4. reasons "for and against"
5. priorities
6 problem-causes-solution
7. cause and effect
8. simple to complex
9. comparison-contrast
10. reasons "for and against"

Master Sheet 54

The Purpose of Classification and Partition

We use partition and classification almost every day. Assume that you are shopping for a refrigerator. If you are mechanically inclined, you will probably begin by thinking about the major parts that make up a refrigerator: storage compartment, cooling element, motor, insulation, and exterior casing. With individual parts identified, you can now ask questions about them to determine the efficiency or quality of each part in different kinds of refrigerators. You have partitioned the refrigerator into its components.

You then shop at five stores and come home with a list of twenty refrigerators that seem to be built from high-quality parts. You now try to make sense out of your list by grouping items according to selected characteristics. First, you divide your list into three classes according to size in cubic feet of capacity: small refrigerators, middle-sized refrigerators, and large refrigerators. But size is not the only criterion. You want economy too, and so you group the refrigerators according to cost. Or you might classify them according to color, weight, and so on, depending on your purpose. Figure 1 illustrates the two kinds of division.

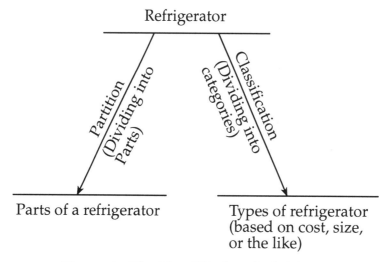

Figure 1   The Two Kinds of Division

These two kinds of division can also be used for more abstract things. You might partition a day into daytime and nighttime or into morning, afternoon, and evening. Or, for other purposes, you may want to classify days, sorting them out as good days and bad days, or profitable days and unprofitable days, and so on.

Partition always deals with one object. Its purpose is to systematically separate that whole object into its parts, pieces, or sections. Classification always deals with an assortment of objects that have some similarities. Its purpose is to group these objects systematically.

Master Sheet 55

Whether you choose to apply partition or classification can depend on your purpose. An architect called upon to design a library will think almost entirely of partition. Once she has defined the large enclosed area that is needed, she must identify the parts into which that space must be divided: the reference area, reading areas, storage areas, checkout facilities, and office space. In some kinds of libraries she might consider providing space for special groups of users (such as reading areas for children). In very large libraries she might need to carry the division further into specialized kinds of space (such as highly secure areas for rare manuscripts or special collections, or areas with special acoustic provisions for listening to recorded materials). But however simple or complex her problem, she is thinking now only about the appropriate division of *space*. She does not have to worry about how the library will classify its books and other material.

But classification is one of the library staff's main problems. The purpose of a library is not only to store books and other forms of information, but above all to make it retrievable. To allow us to find a book or item, the thousands or millions of books stored in the library must be arranged in logical categories. That arrangement becomes possible only if the books are carefully classified.

Master Sheet 56

Suggested Response for Collaborative Project 2, Chapter 12

A. Description of the Cumberland Plateau

    1. Location of the Region

    2. Geological Formation of the Region

    3. Natural Resources of the Region

B. Description of the Strip-Mining Process

    1. Types of Strip Mining

        a. Open-Pit Mining

        b. Auger Mining

        c. Contour Mining

    2. Method of Strip Mining Used in the Cumberland Plateau

C. Environmental Effects of Strip Mining

    1. Permanent Land Damage

    2. Increased Erosion

    3. Water Pollution

    4. Increased Flood Hazards

D. Economic and Social Effects of Strip Mining

    1. Depopulation

    2. Unemployment

    3. Lack of Educational Progress

Master Sheet 57

Chapter 12 Quiz

Name _____ Section _____

Indicate whether statements 1–6 are TRUE or FALSE by writing *T* or *F* in the blank.

1. _____ Once the writing process has begun, a working outline never should be changed.

2. _____ Classification is always applied to a single item.

3. _____ In technical writing, the topic sentence usually appears first in the paragraph.

4. _____ Short paragraphs never should be used in technical writing.

5. _____ A storyboard is especially useful in preparing a long document.

6. _____ Sentence outlines are used mainly in large projects, with team members writing individual sections of a report.

In 7–10, choose the letter of the expression that best completes each statement.

7. _____ In technical documents, outline notation often takes the form called (a) decimal notation, (b) logarithmic notation, (c) alphanumeric notation, (d) holistic notation, or (e) individuated notation.

8. _____ If you were designing and equipping a health club, your organizing task would involve (a) mainly classification, (b) mainly partition, (c) both partition and classification, (d) holistic segmentation, or (e) insulation.

9. _____ Any division in an outline must yield (a) at least one subpart, (b) at least two subparts, (c) three or more subparts, (d) no subparts, or (e) an equivalent denominator.

10. _____ The sequence of parts in an outline is determined by (a) the users' needs, (b) the logic of the subject itself, (c) both a and b, (d) the writer's preferences, or (e) a, b, and d.

Revising for Readable Style

When they are no longer studying matters of style in a void, students begin to see how the correct choices can help writers connect with their audience—how decisions about what to say must be followed by those about *how* to say it.

Students need to understand that "style," in this context, generally excludes literary or any other devices that make writing "fancy." We worry about style insofar as it helps advance our meaning. Instead of calling attention to itself, good style in professional writing should remain more or less invisible.

For upper-level students, the sections on pretentious language and jargon might cause some debate, for many have read such material in papers and articles in their fields. Stick to your guns on this one: explain that plain English is the best English and that a big difference separates choosing words to impress (which invariably makes us sound pretentious) and choosing words for precision (which enables us to say what we mean—clearly). You might wish to use Master Sheets 58–60 to make your point. (The writing samples are from actual memos in organizations that will remain unnamed.) Ask students to comment on the persona in each sample on Master Sheet 59 and to translate the messages (if possible) into plain English.

Master Sheet 61 supplements the discussion of analogy as a way of explaining complex material. The extended analogy here helps readers understand something unfamiliar (computer programming) by comparison with something more familiar (writing).

Passive and active constructions pose a perennial problem. You may need to refer to this section often during the semester. Because it covers a major stylistic issue, this material seems more appropriately placed early in the text than in the appendix. Similarly, the section on conciseness should be consulted often throughout the semester—especially during work on revisions.

For the section on bias, you might bring in newspaper articles and display them on the opaque projector. Or you might use identical lead articles from *Time*, *Newsweek*, and *U.S. News and World Report* on some current controversy, to compare points of view and identify biased writing in so-called objective reporting.

Finally, emphasize that proofreading and revision are done not only for grammatical corrections, but also to improve the content and organization, and finally to achieve clarity and precision of meaning, thereby avoiding disasters like these:

- In my attempt to kill a bee, I drove through a store window.

- After driving for ten years, I fell asleep and hit a tree.

- It is estimated that only 7 percent of the annual white pine mortality in Northern Idaho is harvested every year.

- Because the truck was all over the road, I had to swerve continually before hitting it.

- The need for increased timber production is economical.

- Take the wheel and lie on the ground.

- After becoming familiar with the specifications for interior paneling, this report will consider the economics involved in processing and paneling.

Emphasize that all assignments must be revised because revision is the key to polished, precise writing.

Suggested responses to the style exercises are on Master Sheets 63–81. For a quick and easy classroom review of student responses to these exercises, you might reproduce these master sheets so that students can evaluate their own (or a classmate's) work. As an alternative, you can show the master sheets on the opaque projector.

An Important Qualifier about Style

Whether teaching in the classroom or consulting in business and industry, you inevitably will encounter a version of this protest from someone in the group:

> All this advice about style is fine, but I have a boss who writes in the most flowery language, and expects employees to do the same.

A realistic response:

> If your employer insists on needless jargon or elaborate phrasing, then you have little choice in the matter. Here we study what is best in matters of style; but what is best is not always what some people consider appropriate. Determine quickly what writing style your employer or organization expects. For short-term necessity, play by the rules; but for long-term practice, remember that most documents that get superior results are written in plain English.

Master Sheet 58

Efficiency and Your Documents

What Is an Efficient System?

The efficiency of any system is the ratio of useful output to input. For the product that comes out, how much energy goes in? In an efficient system, the output nearly equals the input.

ENERGY (input) ⟶ | THE SYSTEM | ⟶ PRODUCT (output)

What Is an Efficient Document?

A document's efficiency is measured by how hard users have to work with the document to get the information they need. Is the product worth the effort?

USER EXPENDS ⟶ | THE DOCUMENT | ⟶ USER OBTAINS
ENERGY (input)   THE INFORMATION
                                                           (output)

In other words, users should not have to spend ten minutes deciphering a message worth only five minutes.

Inefficient messages like those on Master Sheet 59 are a drain on users' energy.

Master Sheet 59

Inefficient Messages

We are presently awaiting an on-site inspection of the designated professional library location by Digital representatives relative to electrical adaptations necessary for the computer installation. Meanwhile, all staff members are asked to respect the off-limits designation of the aforementioned location, as requested, due to the liability insurance provisions in regard to the computer.

The new phone system has proven to be particularly interruptive to the administration office secretaries. It is highly imperative that you take particular note to ensure beyond the shadow of a doubt that all calls are of a business nature as far as possible. We are on a message unit cost system which is probationary at best, and the number of phones in our offices is considerably greater than any of our immediate office neighbors have. N.B.: The internal problem has been particularly vexing with phones not being properly replaced in cradles, constituting an additional dimension to an already perplexing internal telephone system dilemma of the greatest magnitude.

Parents in general want their sons and daughters attending classes on a day-in day-out basis, not, as some may choose, on a staggered basis. It is your responsibility to bring pressure to bear on the students to attend your class and it is also your responsibility to be accountable for not bringing this pressure to bear, if such a situation in point of actual fact exists. In short, if a student can be excessively absent from your class without a reason justifiable to both you and the school administration in regard to this matter and this same student simply passes the course on the basis of having made up the work, then I submit that possibly, in such situations, it could be considered much wiser and of greater economy for the Town of Cobalt to offer a correspondence course in circumstances of this nature since the presence of the teacher and what he or she has to offer in the classroom evidently means little or nothing to both the student and/or the teacher involved.

Master Sheet 60

What Errors Decrease a Document's Efficiency?

- More (or less) information than is needed
- Irrelevant or uninterpreted information
- No clear organization
- More words than are needed
- Fancier or less precise words than are needed
- Uninviting appearance or confusing layout
- No visual aids when people need or expect them

How Are Inefficient Documents Produced?

Inefficient documents are produced by writers who lack a clear sense of *purpose, audience, meaning, organization, or style.* In style matters, for instance, we *think* in plain English but we sometimes forget to *write* that way. We might say to ourselves

I want a better job.

But sometimes instead we write

I desire to upgrade my employment status.

Or, in reporting on a survey of employee attitudes, we might intend to say

When our employees feel inefficient at their jobs, they may lose their commitment.

But sometimes instead we write

The factors that potentially dampen our workers' commitment are those which diminish an employee's sense of job efficiency.

Whatever their cause, inefficient documents make users work too hard.

Master Sheet 61

Translating Jargon into Plain English

This letter, an unfortunate mixture of 65-cent words, jargon, and needless passive constructions, is a version of one published in a newspaper.

In the absence of definitive studies regarding the optimum length of the school day, I can only state my personal opinion based upon observations made by me and upon teacher observations that have been conveyed to me. Considering the length of the present school day, it is my opinion that the day is excessive length-wise for most elementary pupils, certainly for almost all of the primary children.

To find the answer to the problem requires consideration of two ways in which the problem may be viewed. One way focuses upon the needs of the children, while the other focuses upon logistics, transportation, scheduling, and other limits imposed by the educational system. If it is necessary to prioritize these two ideas, it would seem most reasonable to give the first consideration to the primary reason for the very existence of the system, i.e., to meet the educational needs of the children the system is trying to serve.

A plain-English translation:

Although no studies have defined the best length for a school day, my experience and teachers' comments lead me to believe that the school day is too long for most elementary students, especially the primary students.

We can view this problem from the children's point of view (health, psychological welfare, and so on) or from the system's point of view (scheduling, transportation, utilities costs, and so on). But our primary interest is the children, because the system exists to serve their needs.

Master Sheet 62

An Extended Analogy

Like report writers, computer programmers rely on problem-solving skills and procedural thinking to analyze a problem and generate a solution. Both the writing process and the programming process entail a set of deliberate decisions.

Writers and programmers define their process and refine their product. First, they decide about their purpose and their audience's needs. Both must define and express complex thoughts within the constraints of a particular language. They must formulate, organize, and articulate the details of a logical map that allows for no ambiguity and requires absolute precision. Writers revise their reports and programmers debug their programs in a cyclical (recursive) process, continuously refining their meaning, rearranging parts, and sharpening their expression.

Both writers and programmers must generate messages that elicit *one* interpretation only. The writer's instructions, as well as the programmer's, must yield the specified results. Both writers and programmers must be sure of their precise meaning, and must communicate that meaning clearly. Both reports and programs should be written in a style and format that allow readers to follow the line of reasoning clearly, and to focus on the significant material.

To make reports easy for readers to follow, writers provide cues by sensible arrangement (transitions, topic sentences, paragraphs); formatting devices (headings, white space, underlining, varying typeface); and report supplements (title page, table of contents, abstract, appendixes). To make programs easy for users to follow, programmers provide documentation: they explain why the program was developed; they describe what the program will do and explain how it will do it; they give instructions for using the program and interpreting its results.

Written reports and computer programs can have similar problems. Writers may give inappropriate or unclear information, and programmers may include documentation computer users can't follow. Missing or incorrect punctuation may cause the reader or computer to misinterpret a sentence or a program instruction. A misspelled or imprecise word will be rejected by the reader or unrecognized by the computer. And, worst of all, a misleading message (a wrong instruction, something left out, or incorrect data) may cause the reader to act or the computer to run, but with the *wrong* results.

Master Sheet 63

Suggested Responses to Chapter 13 Style Exercises

Revising for Clarity

Sentences to be Revised

Exercise 1. Revise each sentence below to eliminate ambiguities in phrasing, pronoun reference, or punctuation.

a. Call me any evening except Tuesday after 7 o'clock.

b. The benefits of this plan are hard to imagine.

c. I cannot recommend this candidate too highly.

d. Visiting colleagues can be tiring.

e. Janice dislikes working with Claire because she's impatient.

f. Our division needs more effective writers.

g. Tell the reactor operator to evacuate and sound a general alarm.

h. If you don't pass any section of the test, your flying days are over.

i. Dial "10" to deactivate the system and sound the alarm.

Suggested Revisions

a. Call me any evening, but not after seven o'clock on Tuesday. *or:* Call me after seven o'clock any evening but Tuesday.

b. This plan offers countless benefits. *or:* I can't imagine any benefits this plan might offer.

c. This candidate has my highest recommendation. *or:* I can't give this candidate a good recommendation.

d. A visit by colleagues can be tiring. *or:* A visit to colleagues can be tiring.

e. Because Janice is impatient, she dislikes working with Claire. *or:* Because Claire is impatient, Janice dislikes working with her.

f. Our division needs more writers who are effective. *or:* Our division needs writers who are more effective.

g. Tell the operator to evacuate and to sound a general alarm. *or:* After telling the operator to evacuate, sound a general alarm.

h. If you fail any section of the test, your flying days are over. *or:* Unless you pass at least one section of the test, your flying days are over.

i. Dial "10" to deactivate the system, and sound the alarm. *or:* Dial "10" to deactivate the system and to sound the alarm.

Master Sheet 64

Sentences To Be Revised

Exercise 2. Revise each sentence below to replace missing function words or to clarify ambiguous modifiers.

a. Replace main booster rocket seal.

b. The president refused to believe any internal report was inaccurate.

c. Only use this phone in a red alert.

d. After offending our best client, I am deeply annoyed with the new manager.

e. Send memo to programmer requesting explanation.

f. Do not enter test area while contaminated.

Suggested Revisions

a. Replace the main seal on the booster rocket. *or:* Replace the seal on the main booster rocket.

b. The president refused to believe that any internal report was inaccurate.

c. Use this phone in a red alert only. *or:* Use only this phone in a red alert.

d. I am deeply annoyed with the new manager for offending our best client.

e. Send a memo to the programmer who is requesting an explanation. *or:* Send a memo to the programmer to request an explanation.

f. Do not enter the test area while it is contaminated. *or:* Do not enter the test area while you are contaminated.

Master Sheet 65

Sentences To Be Revised

Exercise 3. Revise each sentence below to unstack modifying nouns or to rearrange the word order for clarity and emphasis.

   a. Develop on-line editing system documentation.

   b. We need to develop a unified construction automation design.

   c. Install a hazardous materials dispersion monitor system.

   d. I recommend these management performance improvement incentives.

   e. Our profits have doubled since we automated our assembly line.

   f. Education enables us to recognize excellence and to achieve it.

   g. In all writing, revision is required.

   h. We have a critical need for technical support.

   i. Sarah's job involves fault analysis systems troubleshooting handbook preparation.

Suggested Revisions

   a. Develop documentation for our on-line editing system. *or:* Develop on-line documentation for our editing system.

   b. We need to develop a unified design for construction automation. *or:* We need to develop an automation design for unified construction.

   c. Install a system for monitoring the dispersion of hazardous materials.

   d. I recommend these incentives for improving management performance.

   e. Since we automated our assembly line, our profits have doubled.

   f. Education enables us to recognize and to achieve excellence.

   g. All writing requires revision.

   h. Our need for technical support is critical.

   i. Sarah prepares handbooks for troubleshooting fault analysis systems.

Master Sheet 66

Sentences To Be Revised

Exercise 4. The sentences below are wordy, weak, or evasive because of passive voice. Revise each sentence as a concise, forceful, and direct expression in the active voice, to identify the person or agent performing the action.

- *a.* The evaluation was performed by us.
- *b.* Unless you pay me within three days, my lawyer will be contacted.
- *c.* Hard hats should be worn at all times.
- *d.* It was decided to reject your offer.
- *e.* The manager was kissed.
- *f.* It is believed by us that this contract is faulty.
- *g.* Our test results will be sent to you as soon as verification is completed.

Suggested Revisions

- *a.* We performed the evaluation.
- *b.* Unless you pay me within three weeks, I will contact my lawyer.
- *c.* Wear hard hats at all times.
- *d.* We decided to reject your offer.
- *e.* Three happy clerks kissed the manager.
- *f.* We believe that this contract is faulty.
- *g.* We will send you our test results as soon as we verify them.

Master Sheet 67

Sentences To Be Revised

Exercise 5. The sentences below lack proper emphasis because of active voice. Revise each ineffective active as an appropriate passive, to emphasize the recipient rather than the actor.

a. Joe's company fired him.

b. Someone on the maintenance crew has just discovered a crack in the nuclear-core containment unit.

c. A power surge destroyed more than 2,000 lines of our new applications program.

d. You are paying inadequate attention to worker safety.

e. You are checking temperatures too infrequently.

f. You did a poor job editing this report.

Suggested Revisions

a. Joe has been fired.

b. A crack in the nuclear-core containment unit has just been discovered.

c. More than 2,000 lines of our new applications program have just been destroyed by a power surge.

d. Inadequate attention is being given to worker safety.

e. Temperatures are being checked too infrequently.

f. This report was poorly edited.

Master Sheet 68

Sentences To Be Revised

Exercise 6. Unscramble this overstuffed sentence by making shorter, clearer sentences:

A smoke-filled room causes not only teary eyes and runny noses but also can alter people's hearing and vision, as well as creating dangerous levels of carbon monoxide, especially for people with heart and lung ailments, whose health is particularly threatened by "second-hand" smoke.

Suggested Revision

Besides causing teary eyes and runny noses, a smoke-filled room can alter people's hearing and vision. One of "second-hand" smoke's biggest dangers, however, is high levels of carbon monoxide, a particular health threat to people with heart and lung ailments.

Master Sheet 69

Revising for Conciseness

Sentences To Be Revised

Exercise 7. Revise each wordy sentence below to eliminate needless phrases, redundancy, and needless repetition.

  a. I have admiration for Professor Jones.
  b. Due to the fact that we made the lowest bid, we won the contract.
  c. On previous occasions we have worked together.
  d. She is a person who works hard.
  e. We have completely eliminated the bugs from this program.
  f. This report is the most informative report on the project.
  g. Through mutual cooperation we can achieve our goals.
  h. I am aware of the fact that Sam is trustworthy.
  i. This offer is the most attractive offer I've received.

Suggested Revisions

  a. I admire Professor Jones.
  b. Because we made the lowest bid, we won the contract.
  c. We have worked together.
  d. She works hard.
  e. We have eliminated the bugs from this program.
  f. This is the most informative report on the project.
  g. By cooperating we can achieve our goals.
  h. I know that Sam is trustworthy.
  i. This is the most attractive offer I've received.

Master Sheet 70

Sentences To Be Revised

Exercise 8. Revise each sentence below to eliminate *There* and *It* openers and needless prefaces.

a. There was severe fire damage to the reactor.

b. There are several reasons why Jane left the company.

c. It is essential that we act immediately.

d. It has been reported by Bill that several safety violations have occurred.

e. This letter is to inform you that I am pleased to accept your job offer.

f. The purpose of this report is to update our research findings.

Suggested Revisions

a. The reactor was severely damaged by the fire.

b. Jane left the company for several reasons.

c. We must act immediately.

d. Bill has reported several safety violations.

e. I am pleased to accept your job offer.

f. This report updates our research findings.

Master Sheet 71

Sentences To Be Revised

Exercise 9. Revise each wordy and vague sentence below to eliminate weak verbs.

  a. Our disposal procedure is in conformity with federal standards.

  b. Please make a decision today.

  c. We need to have a discussion about the problem.

  d. I have just come to the realization that I was mistaken.

  e. Your conclusion is in agreement with mine.

  f. This manual gives instructions to end users.

Suggested Revisions

  a. Our disposal procedure conforms with federal standards.

  b. Please decide today.

  c. We need to discuss the problem.

  d. I have just realized I was mistaken.

  e. Your conclusion agrees with mine.

  f. This manual instructs end users.

Master Sheet 72

Sentences To Be Revised

Exercise 10. Revise each sentence below to eliminate needless prepositions and *to be* constructions, and to cure noun addiction.

    a. Igor seems to be ready for a vacation.

    b. Our survey found 46 percent of users to be disappointed.

    c. In the event of system failure, your sounding of the alarm is essential.

    d. These are the recommendations of the chairperson of the committee.

    e. Our acceptance of the offer is a necessity.

    f. Please perform an analysis and make an evaluation of our new system.

    g. A need for your caution exists.

    h. Power surges are associated, in a causative way, with malfunctions of computers.

Suggested Revisions

    a. Igor seems ready for a vacation.

    b. Our survey found 46 percent of users disappointed.

    c. If the system fails, sound the alarm.

    d. The committee chairperson made these recommendations.

    e. We must accept the offer.

    f. Please analyze and evaluate our new system.

    g. Be careful.

    h. Power surges cause computer malfunctions.

Master Sheet 73

Sentences To Be Revised

Exercise 11. Revise each sentence below to eliminate inappropriate negatives, clutter words, and needless qualifiers.

a. Our design must avoid nonconformity with building codes.

b. Never fail to wear protective clothing.

c. Do not accept any bids unless they arrive before May 1.

d. I am not unappreciative of your help.

e. We are currently in the situation of completing our investigation of all aspects of the accident.

f. I appear to have misplaced the contract.

g. Do not accept bids that are not signed.

h. It seems as if I have just wrecked a company car.

Suggested Revisions

a. Our design must conform with building codes.

b. Always wear protective clothing.

c. Accept no bids that arrive after May 1.

d. I appreciate your help.

e. We are completing our investigation of the accident.

f. I misplaced the contract.

j. Accept only signed bids.

h. I have just wrecked a company car.

Master Sheet 74

Sentence Sets To Be Revised

Exercise 12. Combine each set of sentences below into one fluent sentence that provides the requested emphasis.

a. The job offers an attractive salary.

   It demands long work hours.

   Promotions are rapid.

   (Combine for negative emphasis.)

b. The job offers an attractive salary.

   It demands long work hours.

   Promotions are rapid.

   (Combine for positive emphasis.)

c. Our office software is integrated.

   It has an excellent database management program.

   Most impressive is its word-processing capability.

   It has an excellent spreadsheet program.

   (Combine to emphasize the word processor.)

d. Company X gave us the lowest bid.

   Company Y has an excellent reputation.

   (Combine to emphasize Company Y.)

e. Superinsulated homes are energy efficient.

   Superinsulated homes create a danger of indoor air pollution.

   The toxic substances include radon gas and urea formaldehyde.

   (Combine for a negative emphasis.)

f. Computers cannot think for the writer.

   Computers eliminate many mechanical writing tasks.

   They speed the flow of information.

   (Combine to emphasize the first assertion.)

Suggested Revisions

a. Although the job offers an attractive salary and rapid promotions, the hours are long.

b. Although the job demands long hours, it offers an attractive salary and rapid promotions.

c. Our office software integrates excellent database management and spreadsheet programs, but most impressive is its word-processing capability.

d. Although Company X gave us the lowest bid, Company Y has an excellent reputation.

e. Although they are efficient, superinsulated homes create the danger of indoor air pollution by such toxic substances as radon gas and urea formaldehyde.

f. Computers eliminate many mechanical writing tasks and speed the flow of information, but they cannot *think* for the writer.

Master Sheet 75

Revising for Exactness

Sentences To Be Revised

Exercise 13. Revise each sentence below for straightforward and familiar language.

a. May you find luck and success in all endeavors.

b. I suggest you reduce the number of cigarettes you consume.

c. Within the copier, a magnetic-reed switch is utilized as a mode of replacement for the conventional microswitches that were in use on previous models.

d. A good writer is cognizant of how to utilize grammar in a correct fashion.

e. I will endeavor to ascertain the best candidate.

f. In view of the fact that the microscope is defective, we expect a refund of our full purchase expenditure.

g. I wish to upgrade my present employment situation.

Suggested Revisions

a. Good luck.

b. Smoke fewer cigarettes.

c. Within the copier, a magnetic-reed switch replaces the microswitches used on older models.

d. A good writer knows grammar.

e. I will try to identify the best candidate.

f. We expect a full refund for this defective microscope.

g. I want a better job.

Master Sheet 76

Sentences To Be Revised

Exercise 14. Revise each sentence below to eliminate useless jargon and triteness.

  a. For the obtaining of the X-33 word processor, our firm will have to accomplish the disbursement of funds to the amount of $6,000.

  b. To optimize your financial return, prioritize your investment goals.

  c. The use of this product engenders a 50-percent repeat consumer encounter.

  d. We'll have to swallow our pride and admit our mistake.

  e. We wish to welcome all new managers aboard.

  f. Managers who make the grade are those who can take daily pressures in stride.

Suggested Revisions

  a. The X-33 word processor will cost our firm $6,000.

  b. To make the most money, rank your investment goals.

  c. Fifty percent of people using this product are repeat customers.

  d. We'll have to admit our mistake.

  e. We wish to welcome all our new managers.

  f. Successful managers are those who cope with daily pressures.

Master Sheet 77

Sentences To Be Revised

Exercise 15. Revise each sentence below to eliminate euphemism, overstatement, or unsupported generalizations.

   a. I finally must admit that I am an abuser of intoxicating beverages.

   b. I was less than candid.

   c. This employee is poorly motivated.

   d. Most entry-level jobs are boring and dehumanizing.

   e. Clerical jobs offer no opportunity for advancement.

   f. Because of your absence of candor, we can no longer offer you employment.

Suggested Revisions

   a. I am an alcoholic.

   b. I lied.

   c. This employee is lazy.

   d. Some entry-level jobs are boring and dehumanizing.

   e. Some clerical jobs offer limited opportunity for advancement.

   f. Because of your dishonesty, you're fired.

Master Sheet 78

Sentences To Be Revised

Exercise 16. Revise each sentence below to make it more precise or informative.

a. Our outlet does more business than Chicago.

b. Anaerobic fermentation is used in this report.

c. Loan payments are due bimonthly.

d. Your crew damaged a piece of office equipment.

e. His performance was admirable.

f. This thing bothers us.

Suggested Revisions

a. Our outlet does more business than Chicago's.

b. Anaerobic fermentation is discussed in this report.

c. Loan payments are due once every two months.

d. Your construction crew shattered two of our computer screens.

e. His management of the Blue Canyon project was outstanding.

f. This gradual rise in interest rates makes us skeptical about further investment.

Master Sheet 79

Adjusting Tone

Sentences To Be Revised

Exercise 17. The sentences below have inappropriate tone because of pretentious language, unclear expression of attitude, missing contractions, or indirect address. Adjust the tone.

- *a.* Further interviews are a necessity to our ascertaining the most viable candidate.
- *b.* Do not submit the proposal if it is not complete.
- *c.* Employees must submit travel vouchers by May 1.
- *d.* Persons taking this test should use the HELP option whenever they need it.
- *e.* I am not unappreciative of your help.
- *f.* My disapproval is far more than negligible.

Suggested Revisions

- *a.* To identify the best candidate, we need more interviews.
- *b.* Don't submit the proposal if it isn't complete.
- *c.* You must submit travel vouchers by May 1.
- *d.* As you take this test, use the HELP option whenever you need it.
- *e.* I appreciate your help.
- *f.* I strongly disapprove.

Master Sheet 80

Sentences To Be Revised

Exercise 18. These sentences contain too few *I* or *we* constructions or too many passive constructions. Adjust the tone.

   a. Payment will be made as soon as an itemized bill is received.

   b. You will be notified.

   c. Your help is appreciated.

   d. Our reply to your bid will be sent next week.

   e. Your request will be given our consideration.

   f. My opinion of this proposal is affirmative.

   g. This writer would like to be considered for your opening.

Suggested Revisions

   a. We will pay you as soon as we receive an itemized bill.

   b. We will notify you.

   c. I appreciate your help.

   d. We will send our reply to your bid next week.

   e. We will consider your request.

   f. I support this proposal.

   g. Please consider me for your opening.

Master Sheet 81

Sentences To Be Revised

Exercise 19. These sentences suffer from negative emphasis, excessive informality, biased expressions, or offensive usage. Adjust the tone.

a. If you want your workers to like you, show sensitivity to their needs.

b. By not hesitating to act, you prevented my death.

c. The union has won its struggle for a decent wage.

d. The group's spokesman demanded salary increases.

e. Each employee should submit his vacation preferences this week.

f. Each applicant must prove that he or she has received his or her state certification.

g. While the gals played football, the men waved pom-poms.

h. Aggressive management of this risky project will help you avoid failure.

i. The explosion left me blind as a bat for nearly an hour.

j. This dude would be a dynamite employee if only he could learn to chill out.

Suggested Revisions

a. Your sensitivity to workers' needs will go a long way toward improving their attitudes.

b. Your quick action saved my life.

c. The union has won a wage increase.

d. The group's spokesperson demanded salary increases.

e. All employees should submit their vacation preferences this week.

f. Each applicant must show proof of state certification.

g. While the women played football, the men waved pom-poms.

h. Aggressive management of this risky project will increase the chance for success.

i. The explosion left me totally blind for nearly an hour.

j. This person would be an excellent employee if only he could learn to relax.

Master Sheet 82

Chapter 13 Quiz

Name _____ Section _____

Indicate whether statements 1–5 are TRUE or FALSE by writing *T* or *F* in the blank.

1. _____ The passive voice usually is more forceful and direct than the active voice.

2. _____ Whenever possible, you should preface your assertions with "I think," "In my opinion," "I believe," or some other qualifier.

3. _____ You should avoid using short sentences in technical writing.

4. _____ Never use *I* in technical writing.

5. _____ The less specialized your audience, the fewer acronyms you should use.

In 6–10, choose the letter of the expression that best completes each statement.

6. _____ In its style, an efficient sentence is clear, concise, and (a) entertaining, (b) informative, (c) fluent, (d) short, or (e) mellifluous.

7. _____ For best emphasis, *avoid* placing the key word or phrase at the sentence's (a) beginning, (b) middle, (c) end, (d) in the terminal position, or (e) none of these.

8. _____ When combining sentences, place the idea that deserves most emphasis in a clause that is (a) dependent, (b) subordinate, (c) independent, (d) relative, or (e) none of these.

9. _____ Technical communicators generally should avoid (a) analogies, (b) euphemisms, (c) the active voice, (d) short sentences, or (e) topic sentences.

10. _____ For most technical documents, choose a tone that is (a) formal, (b) conversational, (c) serious, (d) embracing, or (e) prosaic.

PART

Visual, Design, and Usability Elements

This section explores the rhetorical implications of graphics, page design, and document supplements, and shows how to enhance a document's access, appeal, visual impact, and usability.

Designing Visuals

Because many students have little understanding of the value and role of visual aids, and little experience translating prose into tables and figures, detailed treatment of this chapter would be time well spent. Advanced students need intensive practice in composing visuals.

None of the visuals covered in this chapter requires specialized drafting skills. Even the poorest artist should be able to compose most of them by hand. For greater precision, ask students working by hand to compose all graphs on graph paper which (if your department budget allows) you might provide. Of course, if your students have access to the appropriate software and hardware, their possibilities for exploring the potential of graphics are almost limitless.

Require at least one visual for each document (except memos or letters), placed as close as possible to the first reference to it in the text. Sometimes a student feels that a visual would work well, but is incapable of producing it (e.g., a photograph or complex line drawing). In these cases, have the student leave space in the report text for that visual (numbered, titled, and labeled), describing, in the blank space, the visual that would appear there.

Students can do exercise 1 at home and bring in the completed work for small-group workshops or for full-class workshops, showing selected samples on the overhead projector.

Be sure to make the point that illustrations should clarify, not clutter, a message. All parts should be labeled clearly and the illustrations should be accompanied by a prose interpretation and discussion. Although visuals are an excellent medium for compressing and organizing data, students need to realize that presenting the reader with raw data is not enough—unless of course that is what the reader wants. The implications of the data must be discussed. Sample documents throughout the text have good visuals, which both complement and are complemented by prose discussion and interpretation.

Exercises 7, 8, and 9 work well in class. For exercise 7, you might ask each student to bring in an example of an effective and an ineffective visual. These can be easily shown on the opaque projector directly from textbooks or journals.

Discussion of Exercise 8

The vertical scale is inappropriate; the visual relationships fail to parallel the actual numerical relationships. The ordinate should begin at 0 with each segment of the graph paper equal to 5 or 10 millimeters of CO and so numbered at intervals of 25 or 50.

Answers to Exercise 9

a. table

b. pie chart

c. multiple-line graph

d. bar graph

e. segmented bar graph

f. photograph

g. flowchart or schematic diagram

h. table

i. multiple-bar graph

j. multiple-line graph

Master Sheets for Showing Responses to Exercises

In addition to the Chapter Quiz, this chapter of the Instructor's Manual includes twelve master sheets to show how visuals are designed for various purposes. If you have a microcomputer demonstration room and the appropriate software, you might demonstrate how visuals are composed on a microcomputer. If not, you can use the master sheets on an opaque projector or reproduce them as transparencies or handouts.

Begin with Master Sheet 83, "A Visual Plan Sheet." This material makes a useful handout for summarizing Chapter 14.

For Exercise 1

- Master Sheet 84 shows a table in response to exercise l(a). A table would be most effective here for readers who need to extract figures.

- Master Sheet 85 shows a line graph (with a prose interpretation) in response to Exercise 1(b).

- Master Sheet 86 shows two possible versions of bar graphs in response to Exercise 1(c)

For Exercise 11

The table shown on textbook page 302 has these deficiencies:

- No units of measurement are given. (It should be percentages.)

- Decimals are not rounded off.

- No column averages are given. In addition, the percentage change from 1970 to 1980 would be useful to readers.

- No prose explanation of the comparisons is given.

Master Sheet 87 shows a revised table, along with its interpretation.

For Exercise 12

- Master Sheet 88 shows two possible responses to exercise 12(a). (To save space, the 1980 data is omitted from the sample response.) The pie charts seem best for showing the complete energy picture for either 1970 or 1990—with the exploded segment emphasizing the increase in sources of nuclear energy.

The bar graph, on the other hand, emphasizes side-by-side comparisons for each source—comparisons that perhaps could suggest the beginning of a trend.

General readers easily could interpret either format. The choice, then, of bar graph or pie chart would depend on the writer's purpose.

- Master Sheet 89 shows two possible responses to exercise 12(b). The table is good for *exactness.* Moreover, readers are accustomed to seeing temperatures expressed in numerical form. These readers have a technical use for the information: they presumably will check the temperatures against threshold temperatures in order to know when air-conditioning and heating systems would have to be employed in each city.

  The high-low chart, on the other hand, is good for *emphasis.* By merely scanning the chart, readers can quickly spot cities with the smallest or greatest temperature variations (e.g., temperatures in Miami vary least; those in Dallas, most).

- Master Sheet 90 shows a horizontal bar graph in response to exercise 12(c). Exact figures presumably would not be needed by the Student Senate. The bar-graph format gives readers a quick way of comparing among and within colleges. The black (and most conspicuous) bars represent the *graduates,* probably the major criterion for this audience. Categories are ranked in descending order of numbers of graduates—adding emphasis to that statistic.

- Master Sheet 91 shows a line graph in response to exercise 12(d). The Student Senate presumably wants to observe and compare enrollment trends. Thus the exact values for any year are likely to be less important than an overall view of enrollments over the decade.

Miscellaneous Visuals

- To emphasize interpretation—what to look for in a visual, and what it means—use Master Sheet 92. Like any complex visual, this overlay graph can be hard to interpret.

- Use Master Sheet 93 to show how the segments in a pie chart can be subdivided for greater detail.

Master Sheet 83

A Visual Plan Sheet

Focusing on Your Purpose

- What is this visual's purpose (to instruct, persuade, create interest)?

- What forms of information (numbers, shapes, words, pictures, symbols) will this visual depict?

- Should the audience focus on one exact value, compare two or more values, or synthesize a range of approximate values?

- What kind of relationships will the visual depict (comparison, cause-effect, connected parts, sequence of steps)?

- What judgment or conclusion or interpretation is being emphasized (that profits have increased, that toxic levels are rising, that X is better than Y, that time is being wasted)?

- Is a visual needed at all?

Focusing on Your Audience

- Is this audience accustomed to interpreting visuals?

- Is this audience interested in exact numbers or an overall view?

- Which type of visual will be most accurate, representative, accessible, and compatible with the type of judgment or action or understanding expected from the audience?

- In place of one complex visual, would two or more simpler ones be preferable?

Focusing on Your Presentation

- What enhancements, if any, will engage audience interest (colors, patterns, legends, labels, varying typefaces, shadowing, enlargement or reduction of some features)?

- What medium—or combination of media—will be most effective for presenting this visual (slides, transparencies, handouts, large-screen monitor, flip chart, report text)?

- To achieve the greatest utility and effect, where in the presentation does this visual belong?

Master Sheet 84

NUMBER OF APPLICANTS FOR COLLEGES X, Y, AND Z
1994–1999

YEAR	COLLEGE X	COLLEGE Y	COLLEGE Z
1994	2,341	3,116	1,807
1995	2,410	3,224	1,784
1996	2,689	2,976	1,929
1997	2,714	2,840	1,992
1998	2,872	2,615	2,112
1999	2,868	2,421	2,267
SIX-YEAR AVERAGE	2,649	2,865	1,982
% CHANGE 1994–1999	+22.5	-28.7	+25.5

This table format serves readers who need exact figures. Such readers are likely to take the time to examine the data closely. Tables are useful for presenting data points, not trends.

The summary statistics (average and percentage change) do, however, allow the reader to extract some trend information: for instance, although Y's applicants averaged higher over the six years, they are on a significant decline.

Master Sheet 85

Chapter 14
Exercise 1(b)

From 1994 through 1999, College Y had the highest number of applicants among the three colleges. Since 1994, however, application figures for Colleges X and Z have struggled upward while applications to College Y have been decreasing. These relationships are displayed below.

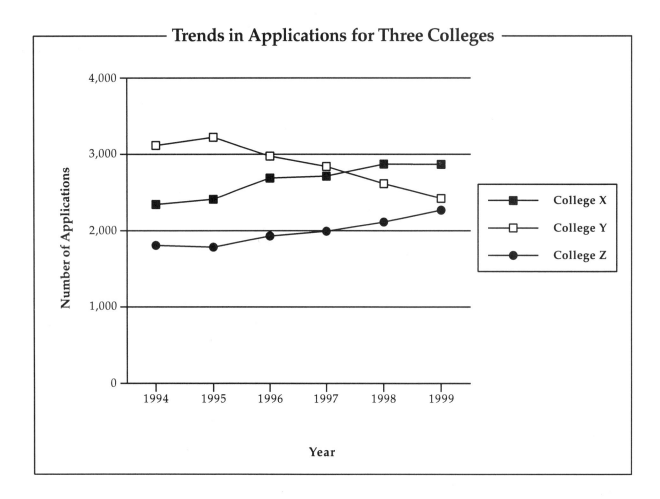

The line graph would be most effective for readers who wonder how overall applications are changing.

Master Sheet 86

Chapter 14
Exercise 1(c)

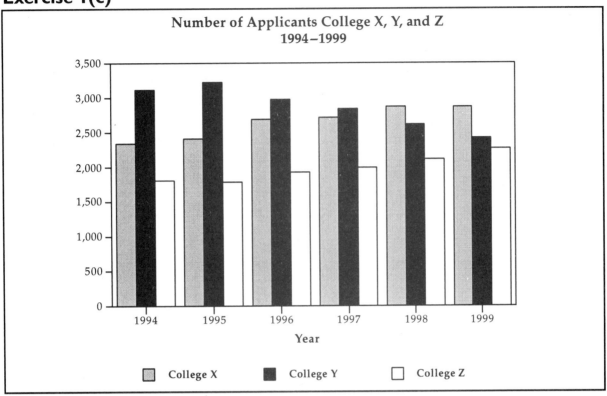

Number of Applicants College X, Y, and Z
1994–1999

Either of these bar charts enables readers to make easy comparisons of application figures for each college within a given year.

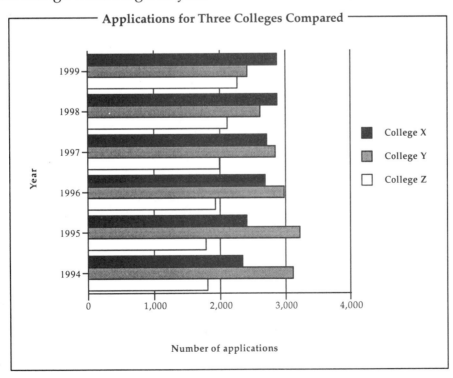

Applications for Three Colleges Compared

Master Sheet 87

Chapter 14
Exercise 11

TABLE 12.4
EDUCATIONAL ATTAINMENT OF PERSONS 25 YEARS AND OVER

HIGHEST LEVEL COMPLETED (IN PERCENTAGES)

YEAR	HIGH SCHOOL (4 YEARS OR MORE)	COLLEGE (4 YEARS OR MORE)
1970	52.3	10.7
1980	66.5	16.3
1982	71.0	17.7
1984	73.3	19.2
1993	80.2	21.9
1996	81.7	23.7
Average	70.8	18.3
%Change 1970–1996	+56.2	+121.4

SOURCE: Adapted from *Statistical Abstract of the United States* (Washington, DC: U.S. Government Printing Office, 1997): 159.

As Table 12.4 indicates, the percentage of persons 25 or older completing high school and college has increased steadily since 1970. The greatest increase, however, occurred among persons completing college, a group whose rate more than doubled between 1970 and 1996.

Master Sheet 88

Chapter 14
Exercise 12(a)

One possible response:

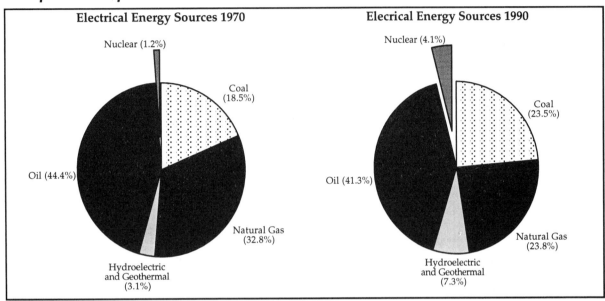

Although coal use shows a slight increase from 1970 to 1990, the most dramatic increase is in the use of nuclear energy. The same period shows a sizable decrease in use of natural gas, and a slight decrease in oil use.

Another possible response:

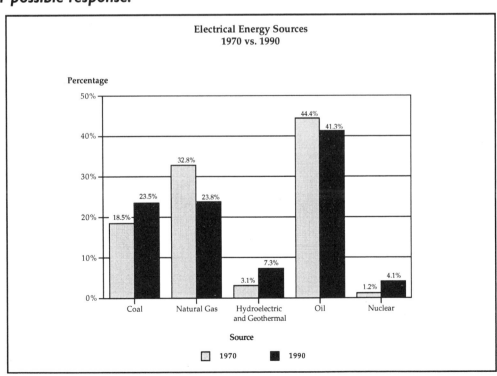

Master Sheet 89

Chapter 14
Exercise 12(b)

One possible response:

TEMPERATURE* AVERAGES FROM 1961 TO 1997 IN MAJOR SUNBELT CITIES

CITY	MAXIMUM	MINIMUM
Jacksonville	78.7	57.2
Miami	82.6	67.8
Atlanta	71.3	51.1
Dallas	76.9	55.0
Houston	77.5	57.4

*Degrees Fahrenheit.

Although average high temperatures for all five cities are near 80 degrees, Atlanta's high temperature is more moderate (nearer 70) and Miami's is higher (nearer 85). Low temperatures, again except for Atlanta and Miami, hover at about 60 degrees. Miami's low is 70, and Atlanta's is 50. Clearly, Miami is at the warm end of the Sunbelt spectrum and air-conditioning costs there are a consideration, but Atlanta's low winter temperatures raise the question of heating expense.

Another possible response:

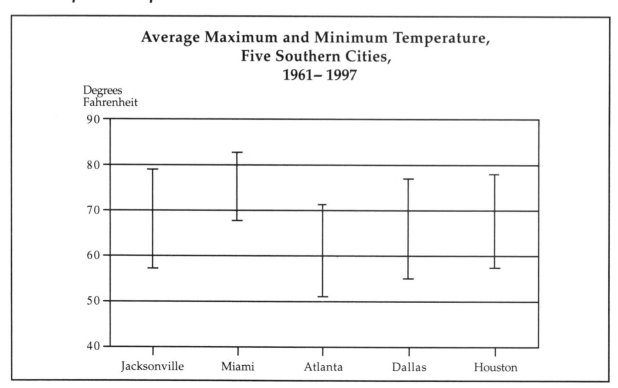

Master Sheet 90

Chapter 14
Exercise 12(c)

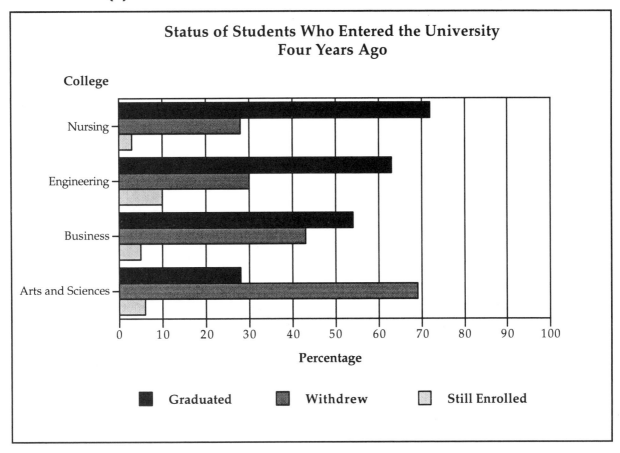

Status of Students Who Entered the University Four Years Ago

College: Nursing, Engineering, Business, Arts and Sciences

Percentage (0–100)

Legend: ■ Graduated ■ Withdrew □ Still Enrolled

The College of Arts and Sciences clearly shows the greatest attrition rate, and Business is graduating just slightly more students than it has lost. In contrast, Nursing and Engineering have retained a high percentage of their students, Nursing having the highest graduation rate.

Master Sheet 91

Chapter 14
Exercise 12(d)

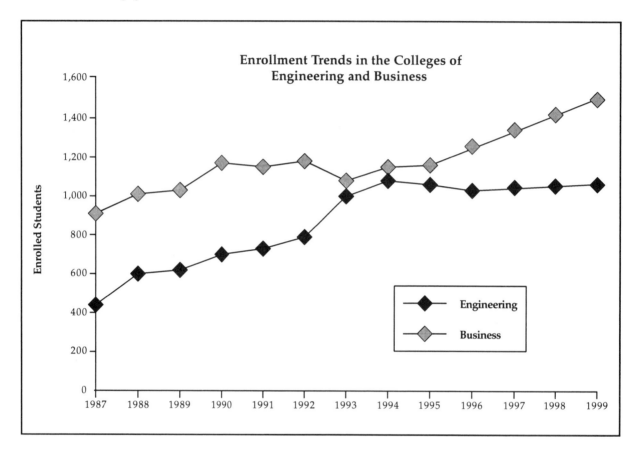

Enrollment trends from 1987 to 1999 show an overall increase in both Engineering and Business. Engineering's sharpest rise in enrollment occurred from 1992 to 1994. Though Business enrollments decreased slightly in 1993, the recent trend suggests steady increase in enrollment.

Master Sheet 92

An Overlay Graph

Overlay graphs combine two or more types of display, usually a bar over which one or more lines appear. The overlay format is good for special emphasis or for plotting two values on the same graph.

Figure X plots two types of values: (1) birth and death rates per 1,000 population, and (2) infant death rates per 1,000 live births.

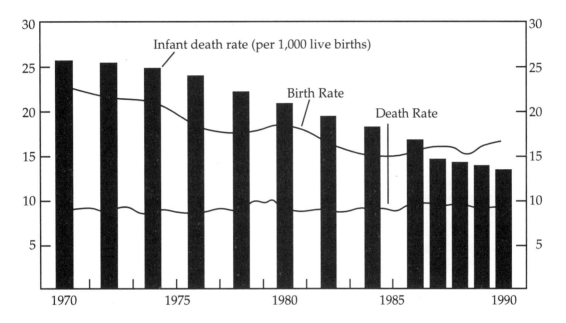

Source: Chart prepared by the U.S. Bureau of the Census

Figure X. An Overlay Graph.

Because overlay graphs can be hard to interpret, use them sparingly, and always explain the relationships depicted:

> Figure X shows a nearly uninterrupted decline in the birth rate from 1970 to 1978, a slight rise to 1980, a continuing decline to 1986, and a gradual rise to 1990. The overall death rate from 1970 to 1990 remained fairly constant, but the infant death rate dropped steadily.

If these trends continue (birth rate rising while infant death rate drops, and overall death rate remains constant), we can expect a growing rate of population increase.

Master Sheet 93

A Subdivided Pie Chart

A pie chart divides a whole into parts. But any of these parts can in turn be subdivided by an additional pie chart. Figure Y shows the original pie chart from textbook page 276, this time with an *exploded* segment that is further divided by a second pie chart.

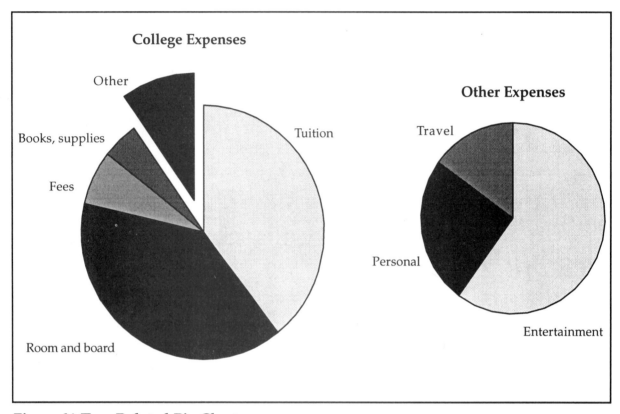

Figure Y. Two Related Pie Charts

Master Sheet 94

Chapter 14 Quiz

Name _____ Section _____

Indicate whether statements 1–5 are TRUE or FALSE by writing *T* or *F* in the blank.

1. _____ Distortion for the sake of emphasis in a visual often is justified.

2. _____ The more complex and richly detailed the visual, the more interesting readers will find it.

3 _____ Pie charts should have no more than seven segments.

4. _____ To illustrate concepts, the parts in a block diagram are represented as symbols or as geometric shapes.

5. _____ For illustrating specific parts in a complex mechanism, photographs are generally better than diagrams.

In 6–10, choose the letter of the expression that best completes each statement.

6. _____ For illustrating a trend, the appropriate figure is typically (a) a line graph, (b) a bar graph, (c) a pie chart, (d) a block diagram, or (e) a photograph.

7. _____ All visuals belong in (a) the report text, (b) appendixes, (c) in either location, depending on their relationship to the discussion, (d) in a glossary, or (e) none of these.

8. _____ A legend is (a) a caption that explains each bar or line in a graph, (b) a prose introduction to a visual, (c) a list that credits data sources for the visual, (d) a visual achievement of historic magnitude, or (e) none of these.

9. _____ To show how the parts of an item are assembled, use (a) an organizational chart, (b) a photograph, (c) a bar graph, (d) a pie chart, or (e) an exploded diagram.

10. _____ An outstanding benefit of computer graphics is (a) ease of use, (b) the endless capacity for "what if" projections, (c) the many colors that can be produced on a map, graph, and so on, (d) clip art, or (e) engaging visual designs.

Designing Pages and Documents

Besides placing formatting power—and responsibility—directly in the writer's hands, automation is raising the audience's expectations of workplace documents. Readers have come to expect documents that look good and that are inviting and accessible.

Many students tend to regard the quality of paper, typing, margins, and the like as annoying trivialities. We therefore need to emphasize that the reader's *first* impression of a document is purely a visual, aesthetic judgment; that a sloppy format is a sure way to alienate the reader; that the time and effort given to "grooming" a report or letter are well spent—and no less important than the time and effort candidates for a job devote to grooming themselves for an interview.

The section on headings deserves detailed treatment, because many students are unfamiliar with their use.

If your institution has an accessible array of microcomputers with word-processing software, have students write drafts on the micros. Tell them to print final hard copy on a letter-quality printer or on a dot-matrix printer that has a near-letter-quality or double-strike mode.

Using the Master Sheets

- Use Master Sheets 95 and 96 to illustrate the influence of format on a document's accessibility and appeal.

- Master Sheet 97 shows the result of excessive highlighting on a document prepared on a microcomputer.

- As your facilities allow, arrange for a demonstration of desktop-publishing software, using a microcomputer and, ideally, a laser printer.

Master Sheet 95

An Ineffective Page Design

Giving the IM Injection

First, the preparation for giving the injection must be carried out. This includes: selecting the correct medication, preparing the needle, and drawing the medication. In selecting the medication, triple-check to ensure that the right medication and dosage are being given by checking the order against the medication card, and, against the label on the drug container. Have the needle ready to go to prevent fumbling with the needle and medication bottle when drawing up the medication. Make sure that the needle is tight to the syringe and that it is the right size. Freeing the plunger so that it will draw back and push forward easily is a good idea because it prevents fighting with it in an awkward position, as in the patient's leg. In drawing up the medication watch several points to avoid contaminating the needle or medication, and to ensure that the right dosage is given. For ease of explanation I assume that the medication is in liquid form, in a container with a rubber seal, and that the dose is correct.

Master Sheet 96

An Effective Design

GIVING THE INTRAMUSCULAR INJECTION

Selecting the Correct Medication and Dosage

CAUTION: Triple-check the physician's order against the medication card and the label on the medication container, to ensure that you administer the correct medication in the precise dosage.

After selecting the correct medication and dosage, prepare your needle and syringe.

Preparing Your Needle and Syringe

1. Choose a twenty-six (26)-gauge needle and affix it tightly to the neck of your syringe.

2. Free the syringe plunger so it draws back and pushes forward easily, to avoid later difficulties when the needle is in the muscle.

With your needle and syringe prepared, you are ready to draw up the medication.

Drawing Up the Medication

CAUTION: Use aseptic technique to avoid contamination of the needle and/or medication; recheck the correct dosage on the container label.

Master Sheet 97

A Design with Excessive Highlights

SUNSPACES

Either as an addition to a home or as an integral part of a new home, sunspaces have gained considerable popularity.

How Sunspaces Work

A sunspace should face within 30 degrees of true south. In the winter, sunlight passes through the windows and warms the darkened surface of a concrete floor, brick wall, water-filled drums, or other storage mass. The concrete, brick, or water absorbs and stores some of the heat until after sunset, when the indoor temperature begins to cool.

The heat **not absorbed** by the storage elements can raise the daytime air temperature inside the sunspace to as high as 100 degrees Fahrenheit. As long as the sun shines, this heat can be circulated into the house by natural air currents or drawn in by a low-horsepower fan.

The Parts of a Sunspace

To be considered a passive solar heating system, any sunspace must consist of these parts:

1. A **collector,** such as a double layer of glass or plastic.

2. An **absorber,** usually the darkened surface of the wall, floor, or water-filled containers inside the sunspace.

3. A **storage mass,** normally concrete, brick, or water, which retains heat after it has been absorbed.

4. A **distribution system,** the means of getting the heat into and around the house (by fans or natural air currents).

5. A **control system** (or heat-regulating device), such as movable insulation, to prevent heat loss from the sunspace at night. Other controls include roof overhangs that block the summer sun, and thermostats that activate fans.

Master Sheet 98

Chapter 15 Quiz

Name _____ Section _____

Indicate whether statements 1–5 are TRUE or FALSE by writing *T* or *F* in the blank.

1. _____ Page design requirements vary from organization to organization.

2. _____ Words in lowercase letters are easier to read than those in uppercase letters.

3. _____ Technical documents usually are read with undivided attention.

4. _____ Margins of 1/2 inch or smaller are desirable for most documents.

5. _____ Whenever possible, begin a sentence after a heading with *This, It,* or some other pronoun that refers to the heading.

In 6–10, choose the letter of the expression that best completes each statement.

6. _____ For highlighting your document, (a) use long lines of italic type, (b) use color generously, (c) use dramatic typefaces sparingly, (d) use FULL CAPS OFTEN, or (e) use all of these.

7. _____ For users who will be facing complex information or difficult steps, you should *not* (a) increase all white space, (b) use a 10-point or smaller type size, (c) shorten the paragraphs, (d) use visuals, or (e) widen the margins.

8. _____ For more personal forms of communication (letters, memos, and so on), choose (a) justified text, (b) unjustified text, (c) short lines, (d) barely justified text, or (e) warm designs.

9. _____ For any document that your audience is likely to read completely (such as a letter, memo, or instructions) you should (a) single-space within paragraphs and double-space between, (b) double-space between all lines, (c) double-space within paragraphs and triple-space between, (d) use long paragraphs, or (e) use any of these.

10. _____ When adding headings, be sure to (a) use no more than two levels of headings, (b) insert one additional line of space above your heading, (c) make each higher-level heading yield at least three lower-level headings, (d) use "catchy" phrasing, or (e) use headings sparingly.

CHAPTER
16

Adding Document Supplements

Much like a good format, well-designed and well-chosen supplements enhance the appeal and accessibility of a longer document. To emphasize the connection between format and supplements in document design, you might assign Chapter 16 in conjunction with Chapter 15.

If you do assign these two chapters simultaneously, be sure students review the Chapter 16 material as they begin drafting their final reports or proposals later in the semester.

As you discuss supplements, remind students that the report text itself almost always embodies a standard organizing pattern: introduction, body, and conclusion—or, orientation, discussion, and review sections. When information is left in its original, unstructured form, readers waste a great deal of time trying to understand and interpret the writer's meaning. The length of each section depends on the relative importance of that section to the report. Instructions, as on textbook pages 479–482, usually begin with a detailed introduction listing materials, equipment, cautions, and so on. The body enumerates each step and substep. Only a brief conclusion follows; the key information was in the procedure itself.

On the other hand, a problem-solving report often has a brief introduction outlining the problem. The body may be quite long, explaining the possible and probable causes of the problem. Because the conclusion includes a summary of findings, an overall interpretation of the evidence, and definite recommendations, it is likely to be detailed. Only when your investigation uncovers one specific answer or one definite cause will the body section be relatively short.

Examples of varying section length, according to subject and purpose, are in the sample reports throughout the textbook.

Answers to Exercise 1

a. This title lacks the word *analysis* or *study;* it suggests that the effectiveness of the system is a foregone conclusion.

b. This title tells nothing about the type of report or its purpose.

c. This title promises too vast a scope. It should at least be limited to positive or negative effects.

d. This title tells nothing about the type of report that follows. Is it a comparative analysis, a description, the history of woodburning stoves?

e. This title tells nothing about the type of report—is it a history, a comparative analysis, or what?

f. This title gives no indication of the report's purpose.

g. Adequate.

h. This title tells nothing about the type of report or its purpose.

i. This title tells nothing about the type of report or its purpose.

Master Sheet 99

Chapter 16 Quiz

Name _____ Section _____

Indicate whether statements 1–5 are TRUE or FALSE by writing *T* or *F* in the blank.

1. _____ Your informative abstract should include your own remarks or opinions, or any new information that you were unable to include in the report itself.

2. _____ Your table of contents should list each heading and subheading included in the report itself.

3. _____ Reports usually are read in linear order, from cover to cover.

4. _____ Four or five appendixes in a ten-page report indicate a poorly organized report.

5. _____ Informative abstracts should be written at the lowest level of technicality.

In 6–10, choose the letter of the expression that best completes each statement.

6. _____ A supplement that precedes the report is the (a) glossary, (b) white space, (c) preview, (d) abstract, or (e) index.

7. _____ The report writer's "off-the-record" remarks or opinions belong in the (a) report abstract, (b) letter of transmittal, (c) appendix, (d) preview, or (e) none of these.

8. _____ Items that do not belong in appendixes include (a) interview questions and responses, (b) recommendations, (c) maps, complex formulas, or (e) interview questions and responses.

9. _____ A letter of transmittal would not (a) discuss the need for follow-up investigations, (b) urge the reader to immediate action, (c) define specialized terms, (d) acknowledge those who helped prepare the document, or (e) refer to sections of special interest.

10. _____ The informative abstract typically (a) aims at highly technical readers, (b) emphasizes only major points in the report, (c) includes any last-minute data that emerged *after* the final draft of the report itself was written, (d) appears as a supplement that follows the report, or (e) does all of these.

Testing the Usability
of Your Document

By focusing on generic usability criteria, this chapter provides a nucleus for classroom workshops in transforming any document from adequate to excellent.

Master Sheet 100

Chapter 17 Quiz

Name _____ Section _____

Indicate whether statements 1–7 are TRUE or FALSE by writing *T* or *F* in the blank.

1. _____ Usability testing is usually performed only on complex products or documents.

2. _____ Online documents tend to be organized in modular fashion.

3. _____ Online documents often are harder to navigate than printed pages.

4. _____ In testing a document for usability, each evaluator should work alone.

5. _____ Usability testing should occur under controlled conditions.

6. _____ Expert users tend to read a document sequentially (page by page.)

7. _____ A basic usability survey focuses primarily on a document's content.

In 8–9, choose the letter of the expression that best completes each statement.

8. _____ Usability criteria for a document are defined by (a) the type of task, (b) user characteristics, (c) constraints of the setting, (d) none of these, or (e) all of these.

9. _____ A usable document enables users to do all of the following *except* (a) easily locate the needed information, (b) understand the information, (c) use the information successfully, (d) develop creative approaches to the task, or (e) carry out the task safely, efficiently, and accurately.

Complete statement 10.

10. Three "human factors" that enhance or limit performance include

_____, _____, and _____.

PART

Specific Documents and Applications

Memos and Electronic Mail

The variety of short reports students will write on the job is far too broad to be covered in one course. As a way of classifying such a diverse collection of short reports, in this chapter we cover two broad forms: letter reports and memorandum reports. Within these two forms are shown the types of reports commonly written: recommendation reports, justification, progress, periodic, and miscellaneous reports. (Short proposals are covered in Chapter 24.)

One problem in discussing short reports lies in their overlapping and often casual nomenclature. For instance, on the job, the designations "progress report," "periodic report," "activity report," or "project report" might be used interchangeably. Students need to know that the concise, highly informative (and often persuasive) writing required in each type of reporting situation is far more important than the "name" of the report.

In a basic course, you might spend a good deal of time on this chapter. Most of the reports here are relatively short and manageable units of discourse, but cover a range of rhetorical situations.

Exercise 4 works well with both basic and advanced groups by providing a hands-on situation and by tying in to the discussion of tone in Chapter 19. Here, as in the complaint letter, the rhetorical purpose is to persuade—not alienate—the reader.

Exercise 7 is useful to both groups, because many students will write minutes at some stage in their careers. Small-group workshops help emphasize the need for precision and for an impartial point of view in reporting. As always, try to do a follow-up, using the opaque projector to show superior examples.

A brief review on the techniques of summarizing will help students achieve the conciseness needed for short reports.

Also, review the material about jargon and letterese in Chapters 13 and 19.

Exercise 11, on appropriate messages for E-mail transmission provides a useful forum for broader discussions about how the medium can influence the message.

Master Sheet 101

Chapter 18 Quiz

Name _____ Section _____

Indicate whether statements 1–6 are TRUE or FALSE by writing *T* or *F* in the blank.

1. _____ As a form of internal correspondence, memos have few legal implications.

2. _____ E-mail increases both the quantity and quality of information.

3. _____ Monitoring of E-mail by an employer is legal.

4. _____ Writing in FULL CAPS increases the readability of E-mail.

5. _____ Short reports are always written at the lowest level of technicality.

6. _____ A memorandum usually should be no more than one page long.

In 7–10, choose the letter of the expression that best completes each statement.

7. _____ Memos are major means of written communication within organizations because they (a) leave a "paper trail," (b) are easy to write and read, (c) are less expensive than other communications media, (d) a and c, or (e) b and c.

8. _____ A typical memo does not have (a) a complimentary close and signature, (b) a subject line, (c) topic headings, (d) a distribution notation, or (e) single spacing.

9. _____ Justification reports are unique in that they are (a) authorized or requested by the readers, (b) often initiated by the writer, (c) rarely written by junior employees, (d) b and c, or (e) a and c.

10. _____ Use your company's E-mail network to send (a) a formal letter to a client, (b) an evaluation of an employee, (c) an announcement for a company picnic, (d) a dinner invitation to a colleague, or (e) any of these.

Letters and
Employment Correspondence

Letter writing assignments (especially job applications) motivate students. In fact, many instructors may prefer to cover this chapter early in the semester as a way of countering resistance to writing. The one drawback to assigning letters early, however, is that students are not yet adept at analyzing the audience, summarizing for conciseness, organizing their material, and displaying a professional format. Until they have mastered the strategies covered in Parts I through IV, students probably will not be ready to produce first-rate letters. One way around this dilemma is to assign a job-application letter very early and then return to letter-writing later. On the second round, students can assess their writing progress by comparing the early letter with later versions. The initial letter provides a good writing sample, and the rewritten letter bolsters confidence in acquired skills.

Some instructors choose to make letters a major emphasis in the course, whereas others treat them only in passing; for that reason, the exercises in this chapter are varied in focus and complexity.

Exercise 1 works well in generating discussion about the specific rhetorical purposes specific letters should serve, and helps solidify the principles discussed in "Elements of an Effective Letter."

Exercise 7 can stimulate class discussions about style and tone. Assign exercise 2 early enough so that students collecting data for research assignments or analytical reports will have time to request and receive the necessary information.

As an alternative to the situations described in exercise 3, you might ask students to write a letter complaining about a product, or one requesting settlement or adjustment of a claim. Good responses to 3 (b) and 3 (c) are shown on Master Sheets 102 and 103. Discuss how each writer makes an *informed* complaint, based on careful study of the issue, and chooses the direct plan (opening with the main point). Emphasize that the primary purpose of a complaint or claim letter is not to tell someone off, but to persuade the reader to act in the writer's favor.

The issue of tone is crucial, and good class discussions can be generated using specific examples with tone problems (some students will invariably submit useful examples).

Tone, in fact, seems to be a major difficulty in letter-writing for students at all levels. The complaint letter is the most dramatic forum for discussions about stylistic objectivity and appropriateness of tone. Discussions of tone related to purpose in complaint letters (as well as in letters of inquiry) will provide good lead-ins to discussions of tone and persona in job-application letters.

Before assigning the job-application letter, you might ask students to evaluate the tone and format in the letters shown on Master Sheets 104 and 105.

Discussion of Master Sheet 104

Besides its ineffective format, the big problem with Raymond Manning's letter is inappropriate tone. Our writer ignores the need for a "you" perspective—especially in a job application. Repeated use of "I" makes the writer seem self-centered. Moreover, the lack of sentence variety and the choppy sentences create a Dick-and-Jane effect, causing the writer to appear unsophisticated, if not simpleminded.

Ask students to revise this letter by trading wordy phrases for single words, eliminating needless prepositions, combining ideas for fluency, and inserting a "you" perspective.

Discussion of Master Sheet 105

Brenda Gaines, in an overstated attempt to convey a tone of self-confidence, comes across as arrogant and condescending. The writer assumes a superior and judgmental posture that is inappropriate, as seen by any reader, especially a prospective employer.

Ask students to revise this letter by adjusting its tone to meet expectations of the situation and the audience.

I have students find a classified ad for a job they could fill once they graduate. The letter and résumé are then written for that definite position. Students attach the ad to their materials for submission. For lower-level students, the Collaborative Project is designed for a heterogeneous group who do not yet have formal job requirements.

With either low- or high-level students, be sure to cover brainstorming while discussing résumé preparation. In fact, you might ask for a class volunteer to perform a public brainstorming session as you record the ideas on the board. The purpose of this exercise is to get students to identify specific qualities and qualifications that make them unique.

Older students with solid work experience should find the résumé on Master Sheets 106–108 a useful model. The application/letter that accompanies that résumé is shown on Master Sheet 109.

An Exercise in Letter Openings and Closings

For a short, informative, and entertaining class exercise, ask students to separate the good from the bad in the openings and closings on Master Sheets 110 and 111. Emphasize the reader's first impression and refer to the Plus-or-Minus column analogy on Master Sheet 8. Whether an otherwise qualified applicant is hired depends strongly on how well the employer *likes* the applicant. The opening should capture the employer's attention, and the closing should move the employer to action.

Discuss the role of *persona*, the impression of the writer that readers derive from the words on the page. Ask students to describe the persona in each example; have them play the role of employer, deciding which applicants they like best.

Discussion of the Letter Openings (Master Sheet 110)

1. The bad joke in the opening (for a job application in law enforcement, no less) trivializes the persona, making the writer ultimately seem uninterested in the job.

2. The self-serving persona here clearly violates the principle that a letter should emphasize what the *applicant* can offer the *employer,* not vice versa. The "you" perspective is nonexistent.

3. This effective opening immediately gives the writer's main point—the U.S. Air Force Academy's *Executive Writing Course* booklet calls it "the one sentence you'd keep if you could keep only one." We get the impression of a writer who is informed, sincere, motivated, and very definite about his or her plans.

4. This opening is familiar and safe, but effective. Nothing flashy—merely a serious, conservative persona, appropriate for this type of job.

5. Busy professionals hate having their time wasted by worthless information. The writer's trite and pathetic attempt at flattery creates a juvenile and unprofessional persona. We get the impression of a person who just can't reach the point.

6. This persona suggests a writer clearly on top of things, informed and action oriented.

7. The main-point lead and the confident but diplomatic tone create a forceful yet likeable persona.

8. The Dick-and-Jane sentence structure and diction, along with a nonexistent "you" perspective, add up to a dreary, faceless persona.

Discussion of the Letter Closings (Master Sheet 111)

1. Being bossy with a prospective employer is no way to create a likeable persona.

2. The humble tone in this example, with its many qualifiers, creates a Milquetoast persona. Nobody likes a wimp.

3. Although definite and determined, this writer oversteps the bounds for an applicant. His "placement" is far from being an established fact.

4. This closing is confident yet diplomatic, both summing up persuasively (assuming the letter text supports such a conclusion) and moving the reader to action.

5. The tone is relaxed and friendly, a good persona for this job.

6. This closing leaves us feeling as if the applicant is a person worth meeting: energetic, motivated, and assertive.

7. This closing reads like something from a set of military orders—a faceless writer telling a faceless reader what to do. All intimacy is missing.

8. This closing effectively sums up the writer's view of her assets, and restates the main point as a diplomatic appeal for action.

An Exercise in Adapting a Message for an Audience

Applications for jobs, grants, graduate school, and so on embody some of the most explicit and challenging forms of persuasive discourse. Now is an ideal time to reinforce students' awareness of the decisions we call the *writing process*—specifically, the centrality of audience awareness in that process.

Comparisons of the first and final drafts of Mary Jo's application letter (Master Sheets 113 and 114) and of Mike's application essay (Master Sheets 116 and 117) stress the notion that a writer's task is at least threefold: (1) to discover one's exact meaning; (2) to communicate that meaning intact; and (3) during the very act of communicating, to refine that meaning, and perhaps to discover new and more significant meanings as well.

By now, students should have had enough practice to appreciate the importance of anticipating their audience's expectations about the shape, style, and substance of a document. Mary Jo's and Mike's first and final drafts (and the accompanying discussions) illustrate the evolution in the writer's discovery and refinement of meaning, and in shaping and expressing that meaning for a stipulated audience.

NOTE: To reduce transparency shuffling, you might project the lead-ins (Master Sheets 112 and 115) and the examples of each first draft (113 and 116), distributing dittoed copies of the final drafts (Master Sheets 114 and 117) for easy comparison and discussion.

Discussion of Mary Jo's First Draft (Master Sheet 113)

1. The inside address is too general, and the salutation is awkward; she should write to a person or at least to a position.

2. The tone in paragraph 1 is plagued by a neutral verb ("observed"); a statement of unwarranted omniscience ("which provides"); a pronoun with a vague referent ("this"), thus obscuring the whole point of the paragraph; and a closing weakened by passive construction.

3. If the reader has not yet discarded the letter, he or she will find paragraph 2—presumably the heart of it—poorly developed, vague ("comfortable atmosphere"), and repetitious ("self-images").

4. Paragraph 3 is neutral and trite ("at a point in my life"), lacks a clear referent ("such an experience") or a near referent ("apply both aspects"), and closes with an awkward and poorly developed sentence.

5. The conclusion is abrupt, weak, trite, and technically inaccurate ("hearing from you"), especially because paragraph 1 merely requested an application.

The persona throughout this letter sounds trivial and vague, and the meaning is unclear.

In contrast, Master Sheet 114 shows Mary Jo's third and final draft. By now, Mary Jo has enhanced the content, overhauled the tone, and organized for a clear line of thought that *shows* instead of merely *telling:* paragraph 1 shows what impressed her, paragraph 2, what her background offers, paragraph 3, what she hopes to gain ("More perfect," an absolute used comparatively, is legitimate in this context); the conclusion is both human and professional.

Most details of what Mary Jo can offer the employer (e.g., résumé items) are absent because her persona is the big issue in this situation. Every bit as vital as résumé items for this kind of job would be the kind of person Mary Jo is.

Discussion of Mike's First Draft (Master Sheet 116)

1. Mike's first paragraph begins with a trite and obvious statement, bound to impress no one. He follows up with an apology that is stilted, wordy, and jargonistic.

2. Paragraph 2 lacks an orienting sentence, is couched in the "passive irresponsible," and is vague, abstract, voiceless, and riddled with needless prepositions.

3. Paragraph 3 continues the earlier paragraph's faulty emphasis on what the writer *expects*, instead of what he has to *offer*. His stated goal, to become "more marketable," is hardly a persuasive reason.

4. Paragraph 4 continues the abstract theme about what the writer hopes to gain, tells readers what they already know, and ends with a vague "This" that drowns the whole point of the essay.

The persona throughout is dreary, faceless, and downright boring, with no "you" perspective and no clear meaning.

Master Sheet 117 shows Mike's fifth and final draft. By now each paragraph begins with a clear orienting sentence, and goes on to offer concrete, persuasive support in a tone that is decisive. His ideas are significant; his plan is clear and sensible; and his attitude is mature, realistic, and engaging. The persona suggests a writer who knows what he wants and how he can contribute.

Schedule several workshops for critiques of the résumé and the job application. Ask students to assume they are personnel managers screening applicants, or that they are helping a friend write a good job application. Have them use the revision checklist as an editing guideline, and ask for detailed commentary and suggestions for revisions. After twenty minutes in a small-group workshop, ask the groups to nominate outstanding pieces to be discussed by the whole class.

Students should revise their letters and résumés according to their editors' comments *before* submitting them to you for final evaluation. Hold another set of small-group workshops during the following period so that the original editor can evaluate the quality of the revision according to her or his earlier comments.

The point is to get students to take an *active* role in discussions about writing—to learn to evaluate, edit, and advise. The more active they are in evaluating someone else's writing, the more discriminating and involved they should become with their own.

Master Sheet 102

A Complaint about a Political Issue

17 River Way
Acushnet, MA 02758
March 3, 20xx

The Honorable Edward King
Governor, Commonwealth of Massachusetts
State House
Boston, Massachusetts 02106

Dear Governor King:

I protest your support of the sale of oil leases on the Georges Bank fishing grounds. As a registered voter of the Commonwealth and a resident of a coastal town, I am convinced that such oil leases would violate the interests of Massachusetts and New England citizens.

In 19xx, New Bedford [a nearby city] was second in the nation in dollar value of all seafood landed. Much of this catch was made up of such prized species as scallops, cod, haddock, flounder, and lobster. This revenue supported much of the local population in fishing and related jobs, such as fish processing and ship repair. Similar situations exist in many of our coastal communities, including Gloucester, Boston, and Provincetown. An industry with this much impact on the state cannot be ignored.

Offshore oil rigs certainly will affect the area's ecology. Sediment, garbage, and oil produced by normal operations on an oil platform will pollute the area surrounding the rigs—an area very close to the scallop and flounder grounds of Georges Bank.

Given the circular water current on the bank, a major blowout or oil spill would not be carried out to sea, but would concentrate on the fishing grounds, thus destroying one of the world's great seafood resources.

The possibility of such a disaster greatly outweighs the benefits from any oil found on the fishing grounds. I therefore ask, in the best interest of the Commonwealth, that you withdraw your support for offshore drilling, and join the citizens who are fighting to prevent it.

Respectfully,

Carol C. Paine

Master Sheet 103

An Effective Claim Letter

18 Second Street
Fall River, MA 02587
5 March 20xx

Mr. Frank Jones
Director of the Physical Plant
Library/Communications Center
Eastern University
Sandwich, MA 02345

Dear Mr. Jones:

Recently, several near-accidents—all within a few feet of the library's main entrance—suggest a critical need for better lighting around the library.

Increased lighting is not a luxury, subject to budget cuts; it is a necessity in preventing accidents and crime. The rising number of thefts and assaults on campus bears out the need for better lighting, not only outside the library, but in all areas of the campus. While the lighting problem exists campuswide, the library (the facility used most at Eastern, especially evenings) seems a logical place to begin.

Everyone is aware of rising electrical costs, but I'm sure you will agree that the University would find a lawsuit more expensive than a few light bulbs. If a student or visitor were to be injured, the University could face a damage suit, not to mention incurring a good deal of bad publicity.

Please install additional exterior lights around the library before a serious misfortune occurs. I make this request on behalf of the many students and faculty who have expressed to me their fear and concern.

Sincerely,

Joseph J. Gutt
Student Representative

Master Sheet 104

An Ineffective Letter of Application

Mr. Arthur Marsh
Durango Chemical Corporation
Box 278
Lakeland, Wisconsin 39765

Dear Mr. Marsh:

I was reading the local paper and came across your advertisement in regards to an opening for a crushing and grinding manager's position at your plant. At the present time I am in college, but would like to fill that opening when this semester is over. I am highly qualified for this job as I have already had two years' experience in this area. I have operated both crushing and grinding circuits that provide the raw ore used in the processing of phosphate products. I also have experience in operating front-end loaders, forklifts, cats, and 30-, 50-, and 120-ton haul units. I have held the different positions of laborer, operator, and foreman, so I have a full understanding of this type of operation. I am a very organized and safety-minded worker who can handle himself well in emergency situations. I am a responsible and punctual employee. If you need any further information concerning my work or personal background, please contact me at the address on the envelope. I thank you for considering my application and hope to hear from you soon.

Cordially,

Raymond Manning

Master Sheet 105

An Ineffective Letter of Application

1289 Fourth Street
Madison, Wisconsin 86743
October 5, 20xx

Mr. James Trask
Trask and Forbes, Attorneys at Law
17 Lord Street
Bartly, Michigan 47659

Dear Mr. Trask:

Having just graduated from law school, I am looking for an established law firm to join. Your firm seems to meet my requirements and I hope I meet yours.

I had thought of going into legal services, but then decided to go immediately into a private practice. I will be able to perform innumerable tasks while gaining invaluable knowledge.

Your firm is considered to be one of the finest in the region and that is another of the aspects that attracted me. Your firm is without a junior partner or assistant at this point in time and I feel very qualified for the position.

Enclosed please find my educational qualifications included in my résumé. I have just passed the bar on my first attempt and received very high grades in law school.

Should you have any questions or comments, we could discuss them at an interview. I am available any time during the business week from 9:00 to 5:00. Feel free to phone me at 304-756-9759 or write, as I would like to hear from you in the immediate future.

Humbly yours,

Brenda Gaines

Master Sheet 106

Résumé from Applicant with Broad Experience

Peter Arthur Profit
14 Cherokee Road • Tucson, Arizona 85703
Telephone: Home 602-516-1234
Office 602-567-5000

QUALIFICATIONS AND CAREER OBJECTIVES

Comptroller, designer of data processing systems, international sales, manager of large foreign office, manager of accounting firm, budget officer.

My immediate goal is to continue my career in fiscal/budget management, in a position with major challenges and responsibilities. Continuing my formal education part-time, I plan to qualify for executive responsibilities.

WORK EXPERIENCE

1995–present *Comptroller*
Datronics, Phoenix, Arizona
Oversee formation of fiscal policies of Datronics; develop appropriate operational procedures; maintain overall coordination of daily business activities, including, for example: (1) supervise development and operation of an accounting system including payrolls, operation, capital equipment budgets, and R&D funds; (2) advise president in forming company policies, plans, and procedures; (3) oversee receipt and control of operational revenues and expenditures; (4) prepare annual budgets and long-range fiscal policy for directors' approval.

1988–1995 *Assistant Manager, then Manager, Financial Operations*
Abernathy's, New York
(1) supervised maintenance of operations accounts; (2) supervised expenditure and receipt of funds (under vice president for operations); (3) developed forms/procedures for accounting, purchasing, cost systems, and computerizing of entire financial operation; (4) established and supervised inventory control system; (5) assisted in preparation of budgets, financial data, and reports.

1983–1988 *Payroll Manager*
Milene's, Boston
(1) supervised payroll department, including preparation of all branch store payrolls, deductions, etc.; (2) responsible for state/federal payroll audits; (3) issued U.S. Savings Bonds; (4) assisted in budget estimates of employee costs and promotions; (5) prepared periodic payroll reports.

Master Sheet 107

1981–1983   *Assistant in Fiscal Management*
Milene's, Boston
(1) supervised three employees in preparation of departmental payrolls and maintenance of relevant personnel records and files; (2) interviewed and recommended applicants for clerical employment; (3) coded and indexed file material.

Note: *At both Abernathy's and Datronics I established the training programs for accounting and computer personnel—programs still in place.*

EDUCATIONAL BACKGROUND

M.B.A. candidate, University of Tucson

Related graduate-level courses: Cases in Personnel Management, Advanced Cost Accounting, Statistical Analysis of Business Trends, Computerized Payroll Systems Development, and others

B.S., Accounting, Northeastern University, Boston, 1981: graduated *cum laude*

A.S., Business Administration, Mass. Bay Community College, Watertown, 1979

Certificate, Proficiency in French, WSAFI, Stuttgart, Germany, 1976

Certificate, Data Processing Specialist, U.S. Air Force Base, Omaha, 1974

PERSONAL INTERESTS, ACTIVITIES, AWARDS, AND SPECIAL SKILLS

Interests: Native American archaeology, skiing, chamber music (I am first violin in an amateur group), gourmet cooking, whitewater canoeing, French and German literature

Activities: 1981 Class Agent, Northeastern University; Rotary Club chapter president (1 year), Framingham, Mass.; United World Federalists chapter treasurer (3 years), Beverly, Mass.; Beverly Hospital Fund chairman (4 years); American Field Service chapter president (3 years), Tucson; United Fund (Commercial) chairman (2 years), Tucson; Sierra Club member (10 years)

Master Sheet 108

Awards: Young Executive of the Year, Beverly Chamber of Commerce, 1987; Record Fund Raising Award, United Fund, Tucson, 1997; Alumni Fund Awards (for highest total), Northeastern University (2 years)

Special Skills: Written and oral fluency—French and German; conversational Spanish; computer operations; successful training programs

REFERENCES

Ms. Janice Stirling Fell, President
Datronics
1142 Arroyo Grande
Phoenix, AZ 85903

Dr. Walter J. Enos, Vice President (Operations)
Milene's
Box 1000
Boston, MA 02114

Ms. Alberta Fresco, President
Abernathy's
500 Fifth Avenue
New York, NY 10014

Mr. Peter S. Pence, Chairman
United Fund
5 Union Place
Tucson, AZ 02103

Master Sheet 109

Letter from an Applicant with Broad Experience

14 Cherokee Road
Tucson, Arizona 85703
November 2, 20xx

Mr. Alfred Wunston, Personnel Director
Arthur D. Lange Company
1213 Massachusetts Avenue
Cambridge, Massachusetts 02138

Dear Mr. Wunston:

I am responding to your call for applications to fill the newly created post of Comptroller—Research and Development, in the January issue of *Aero/Space Journal*. My present position, which I am about to leave, is perhaps somewhat similar, because Datronics, of Tucson, is primarily an R&D organization. In fact, our president, Ms. Janice Fell, knowing Mr. Lange's reputation in the field, has urged me to apply. As you will see from my enclosed résumé, my experience (even to competence in French and German, acquired while stationed in Europe with the U.S. Air Force and routinely used since) may suit your needs.

Having varied experience in most phases of financial management, I feel I could quickly work into your new position. Should the job call for extensive travel and dealing with foreign nationals, I judge myself competent and eager to handle such assignments. My wife and two children also are quite at home in Europe, and they love New England as I do. You may also find of interest my community activities, for they have given me a considerable understanding of environmental and socioeconomic problems—matters with which your company is, I understand, deeply involved. Finally, several times in my life I have had the challenge of building new organizational structures; I would welcome another such opportunity.

Because I am often in Boston on business, I would be happy to discuss this interesting opportunity further. Please call me at home (602-555-1234) or at my office (602-555-5000) any time.

Very truly yours,

Peter Arthur Profit

Master Sheet 110

Letter Openings: Good and Bad

1. After recently paying a parking ticket at the Court House, I noticed an ad on the bulletin board for an opening in the probation department.

2. I recently read of your opening for a field geologist. One of my professors, Dr. R. D. Loner, worked for you, and claims that your company was beneficial to her career. My taking the position would be a great opportunity to advance my career in geology.

3. I have spent many summer vacations hiking and camping in Yosemite National Park, and would like to return to the park as an employee.

4. Please consider my application for the business-oriented programming position advertised in the *Boston Globe*. I will graduate in May from Eastern University, with a B.S. in Electrical Engineering Technology.

5. Texaco is a very important leader in the development and distribution of the world's energy resources. Would your company have a place on its R&D team for a mathematician with experience in computer programming?

6. While attending Eastern University, I have closely followed your company's financial statements, and have become highly interested in your sales growth. Therefore, when Roberta Lowny, Vice-President of Sales for Bando Sportswear, informed me of an opening in your fabric sales division, I decided to write immediately.

7. Does your company have a summer position for a student determined to become a technical writer? If so, I think you will find me qualified.

8. I am applying for a position as a computer clerk. Most of my programming experience has been with PASCAL. I have experience programming in a variety of languages. I was referred to you by Chris Mather, who is employed as a computer clerk in your firm. His interest and enthusiasm encouraged me to write.

Master Sheet 111

Letter Closings: Good and Bad

1. I would like an interview with you as soon as possible.

2. If my qualifications seem to interest you, would you kindly consider the possibility of contacting me at home (247-754-9867) any weekday after 3 P.M.?

3. I know I can succeed as a technical writer in the software industry. I look forward to hearing from you about my placement in your company.

4. I hope you agree that I am the type of engineer DGH is seeking. Please allow me to further discuss career opportunities with you.

5. I feel well qualified for the position as Park Naturalist, and hope you will consider me for the job. If you need more information or wish to contact me for any reason, feel free to call at 265-446-5467 any weekday after 4 P.M.

6. If your department has a place for an enthusiastic intern, I hope you will consider me. I would welcome the opportunity to discuss my application with you directly.

7. I would like to arrange an interview with your company to discuss this position. Please phone me at your convenience.

8. I am hardworking, efficient, eager to learn, and anxious for the opportunity to apply my skills. Please consider me for a summer position.

Master Sheet 112

Mary Jo's Writing Situation

Mary Jo Mooney graduates in three months with a degree in elementary education (language arts). After a winter trip to the Virgin Islands, Mary Jo decides to request an application for a teaching job there. She wants her letter to make a strong impression. She therefore has to carefully assess what her audience would expect from an unsolicited letter. Here are the kinds of expectations we might anticipate from her audience:

About Content

1. legitimate and insightful reasons for applying (what she can offer, what she hopes to gain)

2. an indication of the kind of intuitive insight and sensitivity an elementary teacher especially ought to have

3. concise information (an unsolicited letter to a busy reader with no tolerance for writing that wastes time)

About Organization

1. an introduction that gets to the point and evokes interest immediately

2. a body section that is easy to scan (such letters receive about 30–40 seconds of attention)

3. a conclusion that encourages definite action on the writer's behalf and makes the reader wish to meet the writer

About Style

1. a tone that is purposeful and enthusiastic

2. fluent sentence structure and precise diction (especially for this job)

3. warmth and sincerity

4. a convincingly likable and human persona (people hire you if they like you)

How well does Mary Jo's following draft (her first) meet these expectations?

Master Sheet 113

Mary Jo's First Draft

Willow Drive
East Orange, MA 02768
February 11, 20xx

Department of Education
Roadtown, Tortola
British Virgin Islands

Dear Sir or Madam:

On a recent trip to Tortola, I observed the unique relationship between the children and the adults of the island. The relationship is one of mutual care and respect, which provides children with positive self-images. Seeing this has made me write to ask that a job application be sent to me.

In June of 20xx, I will graduate from Southeastern Massachusetts University and receive my teaching certificate with a B.A. in elementary education (language arts). In my internship, where I developed and utilized a curriculum, I saw the importance of providing a comfortable atmosphere in which children were able to develop positive self-images through learning.

I am at a point in my life where I am able to take on such an experience where I can apply both aspects of my undergraduate work. In spending time on Tortola, I found it to be an excellent area to continue this work.

I look forward to hearing from you.

Sincerely,

Mary Jo Mooney

Master Sheet 114

Mary Jo's Final Draft

Willow Drive
East Orange, MA 02768
February 23, 20xx

Superintendent of Schools
Roadtown, Tortola
British Virgin Islands

Dear Superintendent:

During a recent visit to Tortola, I was struck by the unique relationship that seems to exist between the children and the adults of the island. The mutual care and respect are obvious, and quite apparently provide children with positive self-images. Because my goal is to teach in precisely this kind of encouraging environment, I am writing to ask for a job application.

In June 20xx, I will graduate from Southeastern Massachusetts University and receive my teaching certificate with a B.A. in elementary education (language arts). During my internship, where I developed and implemented a partial curriculum, I saw that a learning-rich environment is one in which children have confidence in themselves—confidence that allows for the kind of exploration essential to real learning.

I had always hoped to find a place more nearly perfect than where I now live (Cape Cod). Your island's weather and beauty truly appeal to me—but even more appealing is the warmth I felt from the people I met. I am determined to return to Tortola. Nothing would make me happier than to teach there.

If any openings should materialize, I would welcome the chance to discuss specific proposals for meeting the needs of your students.

Sincerely,

Mary Jo Mooney

Master Sheet 115

Mike's Writing Situation

Mike Duval, a junior in marine biology, is applying for a prestigious and highly competitive research fellowship at a leading oceanographic institute. Application requirements include a personal essay. Here are the kinds of expectations we could anticipate from his audience.

About Content

1. a concrete and specific proposal for a research project

2. some *new* and *significant* ideas

3. a summary of the writer's qualifications

4. the expected benefits and results of the project

5. neither too much nor too little information

6. answers to questions about *what, why, how,* and *when*

About Organization

1. a distinct line of thought and a clear and sensible plan, consistent with the best scientific methodology

2. an introduction that offers background and justifies the need for the project

3. a body that outlines the scope, method, and sources for the proposed investigation

4. a conclusion that describes the benefits of the investigation and encourages the reader's support

About Style

1. a decisive tone with no hint of ambivalence

2. at least a suggestion of enthusiasm

3. an efficient style, in which nothing is wasted

How well does Mike's following draft (his first) meet these expectations?

Master Sheet 116

Mike's First Draft

I want to study marine science during the summer at Woods Hole Oceanographic Institute as I hope to add to my background and understanding of the marine environment. Presently, I have been unable to conduct any full-time research projects due to the time factor involved and the responsibilities of a full semester's course load. However, your program is an opportunity to study any aspect of the marine environment largely independent of that time factor, except for summer limitations.

Textbooks cannot develop techniques; they can only present concepts. Therefore, one has to develop these techniques himself by actually doing the thinking, the designing, the manipulating, and the interpreting. Once such skills are perfected, they can be carried on for further application in graduate studies or in job situations. And this is an important aspect of the summer program: mastering skills necessary in any kind of research, and developing that "research frame-of-mind." As I plan to further my education by attending graduate school, I think this program will prove invaluable to me.

However, I would like to work a year or so before entering a graduate program so that I can observe senior researchers and understand the requirements of various positions. In this way, and by completing graduate school, I can become a more marketable person.

My research at Woods Hole could lead to continued work as the topic of my graduate thesis. Presently, I am interested in marine microbes and their interactions with invertebrates such as mollusks or crustaceans. Since little is known about these interactions, much attention must be given to this subject. Recent studies have found that some human pathogens are part of the indigenous bacterial fauna of the oyster and other similar shellfish, and can be introduced into the gastrointestinal system via direct consumption. This increases the need to understand and thus control such vectors of human disease.

Master Sheet 117

Mike's Final Draft

Please consider my application for a summer research fellowship. I am a marine biology major at Southeastern Massachusetts University, interested in marine microbes and their interactions with mollusks and crustaceans. My plan is to attend graduate school, but I wish to work a year or so in the biological sciences before enrolling. In this way, I can recognize weak areas in my understanding of biology, and then take the appropriate graduate courses. After graduate studies, I plan to do research and advocate for the environment.

I haven't yet had a chance to conduct full-time research; however, I am studying the indigenous bacterial fauna of the quahog *(Mercenaria)* as a semester project. I am particularly concerned with pathogenic interactions in shellfish, and their ultimate influence on public health through the transfer of dysentery and viral diseases such as herpes, hepatitis, and polio viruses.

Beyond the obvious prestige and resultant professional benefits of a fellowship at WHOI I would anticipate more subtle and personal rewards. These include exposure to graduate-level work and practical applications of the research process—specifically, the design, implementation, and interpretation of an experiment. My natural curiosity and determination to contribute to scientific understanding would, I think, be an asset to your program.

I enjoy working closely with my professors. For the past year, I have been working in a microbiological laboratory under Dr. Samuel Jennings. He, more than anyone, has helped me recognize my scientific abilities and deficiencies. I try to emulate his ways of thinking, incorporating that logic into my own problem solving. With his encouragement, I have learned to approach biology with the precision of a critical observer, the flexibility of an imaginative scientist, and the curiosity of a perennial student.

I realize I have much to learn. Yet, I know enough to ask the kinds of questions that are socially and biologically significant. Are our methods of assessing microbial contamination in shellfish exacting enough? How can we approach a society that leans more toward reaction than prevention, and persuade its citizens that they must change their life-style in order to save their livelihood? A research fellowship at WHOI might enable me to answer these questions.

Master Sheet 118

Chapter 19, Quiz A

Name_____ Section _____

Choose the letter of the expression that best completes each of statements 1–7.

1. _____ The best means of addressing several people at once is to (a) use the salutation "To Whom It May Concern," (b) use the salutation "Ladies and Gentlemen," (c) eliminate the salutation completely by using an attention line, (d) use the salutation, "Dear People," or (e) say nothing.

2. _____ When other documents accompany your letter, add (a) a postscript, (b) a distribution notation, (c) an enclosure notation, (d) a salutation, or (e) an annunciation.

3. _____ The best way to thank a reader would be to write (a) "Thank you," (b) "I wish to express my gratitude," (c) "Please accept my profound thanks," (d) any of these, or (e) none of these.

4. _____ Show respect for your reader by using (a) a formal tone, (b) a "you" perspective, (c) expensive stationery, (d) addressing him/her as "Sir" or "Madam," or (e) all of these.

5. _____ If you expect your reader to react negatively or to need persuading, use (a) coercion, (b) the direct plan, (c) the soft-sell approach, (d) the indirect plan, or (e) a disinformation approach.

6. _____ Make questions in inquiry letters (a) broad and general, to allow readers flexibility in their responses, (b) long and involved, to stimulate reader interest, (c) brief and focused, to save readers' time and ensure definite responses, (d) a and b, or (e) as detailed and prosaic as possible.

7. _____ When making an arguable claim, begin your letter with (a) a clear statement of your complaint, (b) a neutral statement both parties can agree to, (c) a request for immediate action, (d) an apology, or (e) a challenge.

Indicate whether statements 8–10 are TRUE or FALSE by writing *T* or *F* in the blank.

8. _____ Use postscripts generously in your professional letters.

9. _____ Achieve a professional tone by using "letterese" in your correspondence.

10. _____ The subject line is a good device for attracting a busy reader's attention.

Master Sheet 119

Chapter 19, Quiz B

Name _____ Section _____

Choose the letter of the expression that best completes each of these statements.

1. _____ To increase your employment chances, (a) apply for the broadest possible range of jobs, (b) concentrate on selected jobs, (c) have your application letter and résumé written by a professional, (d) a and c, or (e) b and c.

2. _____ The major implied question posed by all employers is (a) why do you wish to work here? (b) what do you have to offer? (c) where would you like to be in ten years? (d) what are your long-term goals? or (e) what salary would you accept?

3. _____ Most employers will scan a résumé in (a) five minutes or less, (b) sixty seconds or less, (c) two minutes or less, (d) four minutes or less, or (e) three minutes or less.

4. _____ Throughout your résumé, use (a) complete sentences, (b) abbreviations, (c) FULL CAPS, (d) passive constructions, or (e) action verbs.

5. _____ Information about religion, color, age, marital status or national origin (a) always should be included in a résumé, (b) never should be included, (c) should be included in specific cases, (d) is required by some employers, or (e) a and d.

6. _____ Generally, the most persuasive references are from (a) relatives and church leaders, (b) professors and previous employers, (c) fellow employees and close friends, (d) intimate others, or (e) employment agencies.

7. _____ Intelligent job seekers (a) limit their search to advertised openings, (b) send some unsolicited applications, (c) phone to request an interview *before* wasting time on unsolicited applications, (d) rely on employment agencies, or (e) use the Internet as their primary search tool.

8. _____ In an unsolicited letter, the opening "I am writing to inquire about the possibility of obtaining a position with your company" is (a) boring and lifeless, (b) conservative and appropriate, (c) the opening preferred by most companies, (d) b and c, or (e) artfully engaging.

9. _____ During an employment interview, (a) you should always have something to say, so as to avoid embarrassing silences, (b) you should realize that the questions you ask are just as important as the answers you give, (c) you never should admit that your knowledge about something is limited, (d) a and b, or (e) b and c.

10. _____ Within a few days of your interview, (a) phone the employer to inquire about your status, (b) send a follow-up letter that restates your interest, (c) relax and wait for the phone to ring, (d) drop in for a surprise visit, or (e) a and b.

Web Pages and
Other Electronic Documents

Following a basic introduction to online documentation and hypertext, this chapter presents the rhetorical features of Web page design, along with brief discussions of HTML scripting language and privacy issues on the Web.

Exercises focus on evaluation of various sites.

Master Sheet 120

Chapter 20 Quiz

Name _____ Section _____

Indicate whether statements 1–7 are TRUE or FALSE by writing *T* or *F* in the blank.

1. _____ Users expect Web documents to offer a traditional introduction, discussion, and conclusion—in the same way a perp document is organized.

2. _____ For effective Web page design, the more graphs and special effects, the better.

3. _____ Online text should ordinarily be at least 50 percent shorter than its hard copy equivalent.

4. _____ Web pages usually take longer to read than paper documents.

5. _____ Persuasive promotional writing increases a Web site's usability.

6. _____ Web site visitors tend to prefer as many navigational choices as possible.

7. _____ HTML is a computer scripting language that can be understood by all Web browsers.

In 8–10, choose the letter of the expression that best completes the statement.

8. _____ Documents on the World Wide Web offer all the following advantages except (a) interactivity, (b) reciprocal use, (c) porousness, (d) continuous and rapid updating, or (e) more rapid reading time than for paper documents.

9. _____ The name for the automated system that allows users to explore a topic from countless angles and to various depths is (a) online documentation, (b) electronic mail, (c) hypertext, (d) fiber optics, or (e) electromagnetic stimulation.

10. _____ Instructions in computer-based training are provided by (a) FAX network, (b) online documentation, (c) telecommuting, (d) boilerplate, (e) browsers.

Technical Definitions

Emphasize the distinction between specialized terms that are overtly technical and more general and familiar terms whose meanings people think they understand.

Discussions about definition provide a good forum for reviewing awareness of the audience information needs covered in Chapter 3. How much and how often one defines will depend on how one views the readers and their information needs.

To avoid problems with plagiarism or with copying from one source, you might require a minimum of four to six references for the expanded definition students choose. With lower-level groups, you might wish to spend one period with the class in the library, pointing out reference books and answering questions that arise during their brief research exercise.

At this time in the semester, students working on analytical reports due at semester's end should have a pretty definite idea of their final topics, after consultation with you. Therefore, the term they choose to define can often be a primary term in that report, such as a definition of biological insect control for the report "The Feasibility of Using Biological Control on Bark Beetles."

Exercise 4 should be done at home and brought in for small-group workshops using the revision checklist. You can then display superior examples via the opaque projector, for further criticism and discussion. Have students revise their expanded definitions at home based on their editor's comments—before you see them—and submit them for your commentary.

Answers to Exercise 1

a. not adequately differentiated from other two-wheeled vehicles such as bicycles or scooters (a two-wheeled vehicle with a pedal-chain configuration to turn the wheel, propelled by the manually driven action supplied by the pumping force of the person pedaling).

b. not precisely classified (electronic device used to control electrical flow); differentiation is circular.

c. "is when" is not an adequate term of differentiation.

d. not adequately classified (an infectious disease of the lymph nodes).

e. neither precisely classified (deodorants and hair sprays are also chemical aerosols) nor adequately differentiated (people other than police, such as national-guard troops or store or bar owners, may use Mace); Mace is a chemical aerosol irritant used to repel attackers.

f. not classified (an electronic instrument).

g. the definition is more technical than the word being defined.

h. adequate.

i. not adequately classified (as a dried plum).

j. not adequately classified (as a force tangential to the abutting surfaces of two bodies).

k. "what happens" is not an adequate term of classification.

l. "important part" is not an adequate classification, and a frame can form the structure of many other items, such as houses or pictures.

m. "a medical term" is not an adequate classification ("a *disease* characterized by . . .").

n. the classification, "device," needs to be expanded.

o. classification is imprecise and points of differentiation are not objective.

p. classification and differentiation are subjective, not factual.

q. classification is imprecise ("to ponder, contemplate, reflect").

Master Sheet 121

Chapter 21 Quiz

Name _____ Section _____

Indicate whether statements 1–3 are TRUE or FALSE by writing *T* or *F* in the blank.

1. _____ Parenthetical definitions are often synonyms.

2. _____ Circular definitions help clarify technical concepts.

3. _____ Expanded definitions usually belong in report appendixes.

In 4–8, choose the letter of the expression that best completes each statement.

4. _____ Abstract and general terms *(condominium, loan, partnership)* often call for (a) parenthetical definition, (b) sentence definition, (c) expanded definition, (d) all of these, or (e) none of these.

5. _____ Definitions should be (a) judgmental, (b) engaging, (c) impressionistic, (d) eclectic, or (e) objective.

6. _____ The specific strategies of expansion you choose will depend on (a) the amount of space in your report, (b) the needs of your audience, (c) the information you have, (d) the time you have, or (e) none of these.

7. _____ Working definitions often are stated as (a) parenthetical definitions, (b) sentence definitions, (c) expanded definitions, (d) eclectic definitions, or (e) none of these.

8. _____ If your report has many parenthetical or sentence definitions, place them (a) in the introduction, (b) in a glossary, (c) at appropriate places throughout the discussion, (d) in an index, or (e) in the informative abstract.

Respond to 9–10.

9. List the three parts of a sentence definition.

10. Besides etymology, history, and background, list three strategies for expanding definitions.

Technical Descriptions
and Specifications

In this chapter we treat description as a rhetorical strategy, but emphasize a specific document format: the descriptive report. On the job, many students will write descriptions of products and mechanisms, and they should be aware that such descriptions demand format and depth of detail that go beyond the requirements of the ordinary descriptive essay, and that such descriptions must be impartial and precise. The structure of the essay (by definition a personal form) elicits descriptions from an expressive rather than a technical point of view.

Also, because descriptive writing is the most tangible in its rhetorical purpose (describing an intact, concrete subject with specific physical dimensions and visual features), the descriptive report provides a good introduction to the more formal reports students will write later. Here they gain practice in developing a detailed outline, using headings effectively, integrating visuals into their discussions, and assessing a reader's needs, as they describe a mechanism for a specific purpose to a specific reader. Also, they practice generating and communicating clear and precise details. If you plan a format report at course's end, encourage students to describe something that may form part of their analysis—such as the kind of solar heating unit to be included in analyzing the feasibility of using solar heat in a greenhouse.

Be sure that each student describes a subject he or she knows intimately, and identifies the audience and the use that the audience will make of the information. We don't describe simply for the sake of describing; our subject, our intention, and what we know of our readers' needs dictate our direction and the amount of detail we include.

Most students (especially lower levels) initially have trouble generating finite descriptive details. One good classroom exercise for overcoming this problem is a variation of brainstorming. Bring to class some mundane and somewhat complex item, such as a coleus plant or a staple remover or a paper punch. Place the item on a table at the front of the class with a ruler positioned conspicuously nearby. Ask the class to write a short piece, on the spot, describing the item or mechanism to someone who has never seen such a thing. After much sweating and grumbling, most students will produce a short piece that is somewhat disorganized and so general as to be meaningless—except for one or two vivid details. Now ask the class as a group to begin assigning descriptive details to the item. Sooner or later, one of them will think to pick up the ruler and measure specific parts. As the details appear, write them all out on the board. Record everything—even those subjective descriptions such as "pretty" and "ugly." Within ten minutes, you should have enough material to fill your chalkboard. Now, ask the class to weed out the subjective from the objective (see pages 443–446). Next, ask them to classify the objective details by dividing the assortment into groups, according to shared characteristics (for the plant: leaves, stem, potting soil, pot; for the staple remover: prongs, plastic finger grips, spring mechanism). Finally, arrange the various classes of detail in the most logical sequence for description (for the plant: from bottom to top, or vice versa; for the staple remover, from finger grips to plastic exterior to hollow metal prongs, including pointed tips and arms, to the coil-spring extensor mechanism). Now decide as a group on the intended audience: Who is it? Why does he or she need the information (to be able to recognize the plant; to manufacture the staple remover, to understand its function)?

After completing this exercise, students should understand what you mean by descriptive details; they should know how to classify data, how to choose the best descriptive sequence, and how to select the appropriate details to fill the reader's specific needs.

If you do exercise 1, divide the class meetings so that you will have one full class to consider outlines only. Get your students to think about planning and organizing. Ask for two or three volunteers to write their outlines on the board for full-class workshops in development and revision. When students bring in the first draft of their full description, ask that they attach a title page and an outline. Hold small-group editing workshops according to the revision checklist. Have students exchange papers with other majors to create a truly general reading audience. After about twenty minutes of small-group workshops, ask students to write detailed suggestions for revision on the report they're reading. Then place an outstanding example on the overhead projector for full-class discussion. It's a good idea to require a written audience analysis with this exercise.

IMPORTANT GUIDELINES FOR EXERCISE 1: Ask students to describe a finite mechanism in specific ways, not a general mechanism in general ways. Encourage them to select a mechanism they can get their hands on—something they can study, weigh, measure, and take apart. Insist that the level of details be keyed precisely to the intended use by the stated audience. Warn the class about getting off the track and ending up writing instructions for operating the mechanism—a most frequent error in this assignment.

As with earlier assignments, ask students to take their first-draft reports home for revision before they submit them for your evaluation.

Master Sheet 122 is the body section of the bumper jack description, with descriptive sequences identified.

This exercise in some ways recapitulates the Agassiz-Schaler exchange. When Agassiz presented his student, Nathaniel Schaler, with a small fish for study, Schaler asked, "What shall I do?" The response: "Find out what you can without damaging the specimen. When I think that you have done the work I will question you." For full context, see "How Agassiz Taught Schaler," in George G. Locke, William M. Gibson, and George Arms, *Toward Liberal Education* (New York: Rinehart, 1958: 5–8).

Master Sheet 122

An Outline for a Mechanism Description

II.   DESCRIPTION OF PARTS AND THEIR FUNCTION

   A.   The Base

      1. Definition

      2. Shape, dimensions, material

      3. Subparts

         a. stabilizing well   **(spatial sequence—bottom to top)**

         b. metal piece

      4. Function

      5. Relation to adjoining parts

      6. Manner of attachment

   B.   The Shaft

      1. Definition

      2. Shape, dimensions, material   **(spatial sequence—small to large)**

      3. Function

      4. Relation to adjoining parts

      5. Manner of attachment

   C.   The Leverage Mechanism

      1. Definition

      2. Shape, dimensions, material   **(spatial sequence—inside to outside)**

      3. Subparts

         a. the cylinder

         b. lower pawl   **(operational sequence)**

         c. upper pawl

         d. "up-down" lever

      4. Function

      5. Relation to adjoining parts

      6. Manner of attachment

   D.   The Bumper Catch

      1. Definition and relation to adjoining parts

      2. Shape, dimensions, material   **(spatial sequence—distal to proximal)**

      3. Function

      4. Manner of attachment

   E.   The Jack Handle

      1. Definition and function

      2. Shape, dimensions, material   **(spatial sequence—distal to proximal)**

      3. Relation to adjoining parts

      4. Manner of attachment

Master Sheet 123

Chapter 22 Quiz

Name _____ Section _____

Indicate whether statements are TRUE or FALSE by writing *T* or *F* in the blank.

1. _____ The main purpose of technical description is to stimulate consumer interest in products.

2. _____ Words such as *impressive, new, improved, large,* and *better* often are seen in technical descriptions.

3. _____ Users of any technical description need as much information as possible.

4. _____ "Phlebotomy specimen" is a more precise and descriptive way of saying "blood."

5. _____ Any item can be described in many ways.

6. _____ When in doubt about how much to include in a description, one should remember that too many details always are better than too few.

In 7–8, choose the letter of the expression that best completes each statement.

7. _____ The most precise technical descriptions are (a) creative, (b) vividly subjective, (c) visionary, (d) objective, or (e) all of these.

8. _____ The details you select for a description will depend on all these elements *except* (a) your purpose, (b) the intended use of the description, (c) your personal preferences, (d) your user's information needs, or (e) your writing situation.

Respond to 9–10.

9. Besides "What is it?", list three reader questions that typically are answered by a technical description.

10. Besides a spatial sequence, list two possible sequences for describing an item.

CHAPTER
23

Procedures and Processes

In this chapter we treat two types of process explanation: (1) instructions—how to do something, and (2) analysis—how something happens.

Within the time limitations of a one-semester course, most instructors are likely to choose to cover instructions only because they are common to many job-related tasks. Instructions therefore are treated in the initial sections of the chapter.

You may, however, wish to give students options in selecting a type of process explanation. Students who plan to be teachers might choose to write a process analysis explaining how something works or how something happens. Students working on a final formal report might wish to write an analysis they could use in the report (say, an analysis of how dwarf mistletoe infects pine trees, to be included in a formal report that will compare and contrast methods of disease control).

Whichever way you approach the chapter, be sure to stress the differences in purpose, emphasis, and point of view in each of the two types of process explanation:

1. Instructions emphasize the reader's role, and are written in the second person.

2. Analysis emphasizes the process itself, and is written in the third person.

If most of the class will be writing instructions, here is a brief exercise to help them develop audience awareness.

Ask each student to take ten or fifteen minutes to write instructions for making a jelly or peanut butter sandwich. Assume that the reader knows what jelly, knife, and bread are, but has just arrived from a country that has no screw-cap jars and no such food as sandwiches. You can expect several students to chuckle or scoff at the apparent simplemindedness of this assignment, but be persistent. Ask them to give the reader all the information needed to complete this task successfully.

Master Sheet 124 is a sampling of responses written by upper-level students—examples chosen from the best writers in the class. (All italics are mine.)

Comments on the examples from Master Sheet 124 follow. As a class, we decided that the instructions in Example A, taken literally, would yield a giant lump of bread and jelly (the writer ignored the assigned topic!), something in the shape of a grapefruit. The class decided that the gaps and ambiguities in Example B make the instructions useless. In Example C, the writer misses the purpose. After a long and largely irrelevant background discussion, he provides incomplete and misleading instructions.

After collecting, reading, and sorting the responses to this assignment, you can read to the class those that are particularly comical or inappropriate, and ask the class to picture themselves performing the procedure as described. Or you might bring to class a loaf of bread, a knife, and a jar of jelly or peanut butter, and ask for student volunteers to read the instructions while you perform the task as described. Again, by choosing samples written by writers acknowledged as superior, you don't risk damaging egos.

Students find this exercise both entertaining and edifying. You now have an excellent forum for discussing the kinds of problems that arise in instruction writing, such as (1) the writer uses language imprecisely; (2) she or he knows the procedure well but makes too many assumptions about the reader's background and ability to fill in information gaps; and (3) the writer misses the purpose completely, as in Example C.

Useful discussions can follow about connecting with the audience, about the need to accurately estimate the skills and background the reader must have to understand the instructions, for the reader will do *exactly* what she or he is told to do, and about how instructions dictate *immediate* action, leaving no room for unclear or misleading information. This simple exercise will reinforce the point about writing for a specific reader's specific needs.

The good news is that instructions are always written chronologically, and so organization becomes easier than it was for descriptive writing. The bad news is that writing instructions is fraught with the liabilities indicated above. Collaborative Project 2 is a good introduction to the complexities in writing good instructions.

Use the same workshop approach here as suggested in earlier chapters. Have students bring in a completed outline before they write the actual instructions. In all cases, be sure they have revised according to their peers' comments *before* they submit a draft for your evaluation. Require a written audience-and-use analysis with the instructions.

Combining the work in Chapters 22 and 23, you could ask students to prepare a manual for a mechanism they know how to operate. The manual would begin with a mechanism description and move on to instructions for its assembly, operation, or maintenance.

Here is a supplementary list of topics for instruction writing:

- How to Mark Timber Unit Sale Boundaries
- How to Obtain Growth Increment and Age from a Tree
- How to Change Bicycle Brake Shoes
- How to Replace a Bicycle Inner Tube
- How to Operate a Grease Recycler
- How to Prepare Explosives for Detonation with Electric Blasting Caps
- How to Pin a Butterfly

One way to avoid plagiarism is to ask students beforehand to submit a list of two or three procedures they can perform well. Then you can assign instructions for one of these procedures. Another way is to have them rewrite instructions that are too technical or otherwise inadequate for a general reader. They should submit a copy of the original along with their own version.

Master Sheet 124

Three Ways of Making a Peanut-Butter Sandwich

Example A

The only materials needed for making a jelly sandwich are a small jar of jelly, two slices of bread, and a knife. With these ingredients close by, the procedure is:

1. Place both slices of bread *beside each other* on a flat surface, preferably a breadboard.

2. *With the knife in one hand, unscrew the lid* of the jelly jar and scoop a large quantity with your knife.

3. Flow a 1–2 mm coating of jelly *on each bread surface,* being careful not to waste any excess.

4. Join the *two fully-coated* bread slices together so that the jelly *surfaces come in full contact with each other.*

If you are not satisfied with your results, repeat steps 1–4 until the desired form, consistency, and quality have been achieved.

Example B

The materials needed for a peanut butter sandwich are a table knife, one jar of peanut butter, and two slices of bread. Hold one slice of bread in the left hand. With the knife in the right hand, spread an even layer of peanut butter about *1/4 inch thick on the piece of bread.* Then, place the other slice of bread *on top of the peanut butter.*

Example C

A peanut butter sandwich is composed of two or more pieces of bread and peanut butter. Bread is commonly used as the foundation of a sandwich, yet such other staples as crackers may be substituted. Bread can be made from a variety of grains such as rye, wheat, bleached flour, or sourdough. These ingredients alter the color, accounting for dark and white bread. The bread is usually sliced into 1/2" to 1" pieces and spread with peanut butter.

Peanut butter is made from peanuts and other stabilizers, if store bought. A creamy mixture results when the peanuts are mashed, and peanut butter is born. A utensil, usually a knife, is used to spread the peanut butter over the bread slices. The amount used depends on personal taste.

After the bread pieces are sliced and spread with peanut butter, they are placed on top of each other so that peanut butter surfaces are touching. Have a glass of milk to aid swallowing.

Master Sheet 125

Chapter 23 Quiz

Name_____ Section _____

Indicate whether statements 1–4 are TRUE or FALSE by writing *T* or *F* in the blank.

1. _____ To avoid cluttering your instructions, use as few transitional expressions as possible.

2. _____ Above all, make your instructions brief.

3. _____ Because users are often impatient, the introduction to a set of instructions should give very little background.

4. _____ Of all kinds of technical communication, instructions have the most demanding requirements for clarity and phrasing.

In 5–10, choose the letter of the expression that best completes each statement.

5. _____ Instructions emphasize (a) the writer's performance, (b) the user's performance, (c) the subject's performance, (d) all of these, or (e) none of these.

6. _____ Any warnings or cautions should be spelled out (a) in an appendix, (b) just before the respective steps, (c) in the introduction, (d) a, b, or c, or (e) a and c.

7. _____ Phrase all instructions in (a) the active voice and imperative mood, (b) the passive voice and the indicative mood, (c) the active voice and the subjunctive mood, (d) any of these, or (e) none of these.

8. _____ Working definitions of specialized terms for instructions belong (a) in the introduction, (b) just before the related step, (c) in a glossary, (d) none of these, or (e) a and c.

9. _____ Usability criteria for a document are defined by (a) its intended use, (b) human factors, (c) both of these, (d) the user's level of interest, or (e) none of these.

10. _____ Any visual in a set of instructions should be (a) placed in an appendix so as not to interrupt the steps; (b) placed in the introduction, to increase interest; (c) incorporated within the discussion of the related step, for immediate reference, (d) a or b, or (e) a or c.

24

Proposals

Whether your course is basic or accelerated, you should include some practice in proposal writing. As time permits, have students regardless of level—read the entire chapter. I usually give accelerated classes the choice of a formal proposal (based on exercise 6) or a formal report as the final term project.

Although exercise 3 is intended for accelerated classes, it can be adapted to basic classes as well, in their planning of research reports. For students planning less technically oriented research than that proposed on textbook pages 497–499, Master Sheets 126–127 and 128 offer model proposals for researching, respectively, a technical problem and a business management problem.

Exercise 1 is designed for basic students, using a letter format. Class discussion of this exercise is an excellent motivational tool.

If your school employs a grant writer, invite this person to speak to your class about the function of proposals in institutions that rely on outside funding. If someone in your school is developing a new program (in computer technology, medical technology, environmental studies, and so on), you might invite him or her to speak about the planning proposals that helped get the program off the ground.

Exercise 2 is designed primarily for basic students but also works well for evening students who have solid work experience. Class discussion produces some lively responses.

With the exception of the research proposal addressed to you, require a written audience-and-use profile with all proposals.

Possible topics for the long proposals in exercise 4 need sharp and directed focus. Be sure students define clearly their situation, audience, and purpose—especially important in a long, complex proposal. To help them achieve adequate focus, take a topic (a proposal for a campus improvement) and refine it, as in the next exercise.

An Exercise in Refining a Proposal Topic

Master Sheet 129 shows how a typical writer of an unsolicited proposal moves through decisions about *situation, audience, problem*, and *purpose*. You might wish to use some version of the following discussion to accompany the master sheet.

The Proposal Situation: The primary reader has a definite problem or need and has asked you to submit a proposal, or you have identified a problem or need and decided to submit an unsolicited proposal. For either problem, readers expect a realistic plan.

The Proposal Audience: Design the proposal for a primary reader other than your writing instructor. Include a written audience-and-use analysis with the proposal, basing your decisions about details and technicality on that analysis. If your primary audience is more specialized than your secondary audience, design the proposal at a technical level, and adjust the supplements for a lower level of technical understanding (say, for your instructor as secondary reader).

Statement of the Problem: An effective proposal is one that persuades readers to take or support the recommended action. Your proposal, therefore, should be a specific response to a specific problem.

Beyond demonstrating your full understanding of the problem, your proposal should enable the primary readers to understand all aspects of the problem and to evaluate the soundness of the plan.

Statement of Purpose: From your precise focus on the problem grows your statement of purpose. Because it includes a summary of your plan for solving the problem, your statement of purpose in turn guides your decisions about exactly what to include in the text of your proposal.

After a session with this material, ask for a student to volunteer another topic that he or she is working on. Then hold a full-class workshop to refine that topic.

Discussion of Master Sheet 130

Designed for classroom use, this chapter emphasizes *internal* proposals. As a motivator, however, you might want to discuss *external* proposals as well. To give students an idea about the kinds of external proposals solicited by just *one* branch of the U.S. government (the largest proposal customer), use Master Sheet 130 with some version of this commentary:

The U.S. government solicits countless proposals from private businesses for supplying items, materials, methods, and studies. Whatever the government's need, the firm offering the best plan wins the contract. The following solicitations are taken from a booklet titled *Program Solicitation Number 84-1, Small Business Innovation Research Program* (Washington, D.C.: Department of Defense, 1983).

All the topics in the SBIR booklet cover relatively small projects (less than $500,000). The guidelines for writing the proposal are included in the booklet. For major projects (such as developing a space-shuttle navigation system), the RFP would be much more detailed than the solicitations shown here.

Although these topics are all defense related, other government departments publish their own solicitations as well.

This is a good time to mention that *creativity* can be a vital element in technical writing. In a proposal, not only must the plan itself be creative, but so too must the proposal writers. They must decide how to present the plan in its best possible light so that it can be fully appreciated by readers.

Master Sheet 126

One Student's Research Proposal

March 20, 20xx

TO: Dr. J. M. Lannon, Director of UMD's Writing Program

FROM: Christina Trinchero

SUBJECT: Proposal for Investigating the Erosion Problem on Cape Cod

Introduction

In 1889, Mr. Henry Marindin of the U.S. Coast and Geodetic Survey conducted a study to determine by how much Cape Cod was retreating into the vast ocean. [Early investigators had estimated that the Cape had originally extended from 2.5 to 4 miles to the east (Leatherman, i)]. Mr. Marindin had set up markers from the Nauset spit to the Provincetown hook, and in 1950, 74 of his original markers were replaced with concrete benchmarks. A research team from Woods Hole Oceanographic Institute determined that over this 70-year period, the shoreline had eroded at an average rate of 2.5 feet per year (Leatherman, i).

Statement of Problem

More than one hundred years after Mr. Marindin's study, the tide still ebbs and flows along the same stretch of Cape Cod beaches. But how much of the original beach remains? How long will waves crash against sandy shores? Will there come a day when Cape Cod is underwater?

Proposed Study

Humans struggle with the natural world constantly. We have been, and probably always will be, in conflict with the natural world. Using technology, humans have found many ways to tame nature, yet some aspects of nature remain untamable. The ocean has yet to be managed by humans. Erosion, the wearing away and removal of any kind of earth materials by water and wind, as scientists have discovered, is unstoppable. It is crucial to detect where erosion has occurred and, if possible, slow the process, saving the land that remains. Such is the case on Cape Cod. Erosion continues to shape and reshape the Cape. The way to address the erosion problem is to preserve the beaches and land the tides have not yet taken out to sea.

Master Sheet 127

<u>Scope</u>

To investigate Cape Cod's erosion problem, I plan to pursue the following areas of inquiry:

- How does erosion occur?
- How does the Cape rebuild itself?
- How did Thoreau describe the Cape when he first saw the area?
- Who was Henry Beston? How does he relate to the erosion issue?
- Is erosion a threat to the Cape's wildlife and plant life?
- What have previous studies revealed about the erosion problem?
- What can be done to help slow erosion on the Cape?
- What is the function of Cape Cod National Seashore?
- What is the National Seashore doing to slow erosion?
- Does building still occur high atop dunes despite warnings of high tides and erosion?
- How do owners of beachfront property feel about erosion? Worried enough, say, to move their home 100–1000 feet back from the shifting sands?
- What can concerned residents of the Cape do to help prevent erosion?

<u>Methods</u>

My primary data sources will include interviews with Mr. John Smith, Director of Environmental Planning at the National Seashore in South Welfleet, MA, and Dr. George Brown, an erosion expert at Woods Hole Oceanographic Institute. I will compare (via photographs) several beaches damaged by erosion and look at an experimental beach designed by the National Seashore to help prevent erosion. I have considered doing a random telephone survey to determine if, and to what extent, Cape Cod residents feel threatened by erosion.

My secondary sources will include two reports: the first is a report on the historical cliff erosion of outer Cape Cod; the second study tells of the interaction of vegetation and geological processes on Barrier Beaches, and off-road vehicle impact on dunes of the Cape Cod National Seashore. I will also consult articles dealing with the erosion issue.

<u>My Qualifications</u>

My only qualification is that I am an avid beach-goer who has an interest in why erosion occurs and in what can be done to prevent it from destroying the Cape.

<u>Conclusion</u>

Clearly, natural cycles can't be stopped; to end a cycle would eventually threaten even the most advanced forms of life. The Cape Cod National Seashore is working to study erosion and determine how to control the natural process. The question remains: is Cape Cod disappearing? With your approval, along with any suggestions, I will continue my research.

Master Sheet 128

Another Student's Research Proposal

March 4, 20xx

TO: Dr. James Granger, Instructor

FROM: Anne Bickett

SUBJECT: ANALYTICAL REPORT PROPOSAL

In response to your March 2 request for an analytical report due May 10, I propose to study the problem of low employee morale at the Foodstuff Supermarket in South Dennis. During my two years there as a part-time employee I have seen high employee turnover and general dissatisfaction over working conditions and management policies. Top management has also voiced its concern about the morale problem.

My analysis, written for management and staff, will cover these areas:

1. an assessment of the efficiency of Foodstuff's management by drawing parallels and contrasts between the actual management operation and principles of management that I will identify and collect from my secondary sources.

2. a consideration of the direct and indirect effects that any management problems might have on employee morale.

3. an assessment of management's effectiveness in employee motivation.

4. an assessment of management-employee communication, and vice versa.

My secondary research will include library and other published sources of data. Primary research will include personal observation, questionnaires, and interviews with both employee and management representatives.

Any recommendations for improvement will be based on the collected evidence in my report. I will be happy to discuss this proposal further with you at any time.

Master Sheet 129

How To Refine a Proposal Topic

The Proposal Situation

Assume you are a computer-science major who has tutored in your school's computer center for two years. You've noticed that less and less terminal time is available to the growing number of users. You decide to address this problem by developing a plan for fair distribution of terminal hours among all users.

The Proposal Audience

The primary audience for your unsolicited proposal will be the dean of faculty and the academic deans (the decision makers). The secondary audience will be the computer-center staff (those who would implement the plan). For your primary audience (non-specialists), you decide to keep your proposal text at a low level of technicality, defining in detail specialized terms such as *on-line registration* and *enforcement protocols.* For the secondary audience (specialists), you include appendixes describing your analysis of the computer system and outlining the technical details of your plan (such as a program you've written for the computer to log out users automatically after sixty minutes).

Your Statement of the Problem

With the growing number of computer-science courses, available terminals on our VAX 2060 have all but disappeared. This critical shortage of terminal time seems to have several related causes:

- abuse (often unintentional) of terminal time by users
- too few terminals to meet demand
- lack of project coordination among faculty (i.e., little effort to distribute projects during a semester)
- inadequate preparation for new users from introductory courses (who can spend up to two hours just trying to "LOGON" to the system)

Your Statement of Purpose

To ensure fairness to all users, and to minimize the waste of terminal time, I propose a Terminal Scheduling System (TSS). The TSS would allow each user to register for a designated number of hours each week, based on the history of required terminal time for the user's course and the number of courses being taken. The TSS could be implemented as a manual or automated system (with sign-up sheets or on-line registration and enforcement protocols).

Master Sheet 130

Typical Proposal Solicitations (SBIR)*

(For an item)

TITLE: *Collapsible Food-Service Bowls*

DESCRIPTION: Develop collapsible bowls that are sturdy, impermeable to water and oil, extendible to one-half liter capacity (slightly more than one pint). These are needed for serving stews and casseroles in remote sites.

(For material)

TITLE: *New Lightweight, High-Tensile, Durable Small-Tent Fabric*

DESCRIPTION: The U.S. Army has a need for one-person and two-person tents that are person-portable, lighter, and smaller than those currently used. New tent fabrics are needed to meet these requirements as well as to provide protection against the effects of high altitude, cold weather, and light penetration.

(For a mechanism)

TITLE: *Robotic Deck Scrubber*

DESCRIPTION: Design and construct a robotic deck scrubber to be used in the hangar deck of a carrier to clean up oil spills. The device would be programmable, with obstacle avoidance sensors. The device would dispense detergents and contain built-in brushes and a vacuum system.

(For a method)

TITLE: *Testing Procedures for Asbestos in Military Facilities*

DESCRIPTION: The Toxic Substances Control Act (TOSCA) prohibits use of asbestos in new buildings, and requires that old buildings such as schools be inspected for asbestos. There is a need for asbestos identification techniques that can be used in the field by technician-type personnel. The identification techniques should be specific enough to satisfy the requirements of TOSCA and be acceptable to the EPA.

(For a study)

TITLE: *Visual Information Processing*

DESCRIPTION: The goal of this basic research program is to develop a quantitative description of human visual processing. Special emphasis is on those aspects of visual processing which are most relevant to air crew performance, selection, and training; rapid and accurate interpretation of visually displayed information; and the development of robotic visual systems.

---

*Small Business Innovation Research Program

Master Sheet 131

Chapter 24 Quiz

Name _____ Section _____

Indicate whether statements 1–5 are TRUE or FALSE by writing *T* or *F* in the blank.

1. _____ Generally, unsolicited proposals have longer introductions than those that have been solicited.

2. _____ Business and government proposals are most often unsolicited.

3. _____ The body section of a proposal receives most attention from readers.

4. _____ In some long proposals, the conclusion can be omitted.

5. _____ Proposals usually are written at the lowest level of technicality.

In 6–10, choose the letter of the expression that best completes each statement.

6. _____ The guidelines for developing a proposal for a specific client typically are included in (a) the RFP, (b) the proposing firm's operations manuals, (c) any comprehensive textbook, (d) b and c, or (e) the letter of transmittal.

7. _____ A proposal for improving your company's employee morale, requested by the vice-president in charge of personnel, probably would be classified as (a) a solicited, internal research proposal, (b) an unsolicited, external planning proposal, (c) a solicited, internal planning proposal, (d) a sales proposal, or (e) none of these.

8. _____ Especially important to nonspecialized audiences for a long proposal is/are (a) the conclusion, (b) the appendixes, (c) the introduction, (d) the abstract, or (e) the letter of transmittal.

9. _____ Successful proposals are usually those that are (a) most specific, (b) most ambitious, (c) most elaborately designed, (d) most creative, or (e) most prosaic.

10. _____ Besides being clear, the proposal plan must be (a) inexpensive, (b) highly optimistic, (c) coercive, (d) creative, or (e) realistic.

Analytical Reports

This chapter culminates the course, especially for upper-level students—although some instructors ask basic classes for a formal report as well. Students should have selected their individual topics early in the semester (see Syllabus B), and ideally will have keyed many of their earlier assignments (definition, description, partition and classification, process narration) to this final report.

Early in the semester, go over in class the report samples in this chapter and in Appendix C to give students a sense of direction and purpose and show them exactly what they are working toward as a semester goal. Ask them to read "Typical Analytical Problems" to identify their specific purpose and approach. Emphasize that the common purpose in all such analyses is to come up with specific recommendations.

Begin the actual work on the analytical report early enough in the semester to give students a chance for one revision according to your comments. Because many students are intimidated by the prospect of this formal report, the revision becomes crucial in improving the quality of their final product.

Encourage students to choose practical topics. Review their proposal to see that they ground their analysis in a specific situation for a specific purpose and audience. Require a report that readers can use in the workplace, derived as much from primary research data as possible.

Discuss at length how this final writing assignment differs from the earlier pieces. Emphasize that earlier assignments almost always dealt with subjects having tangible limits and structures—description, process explanation, classification—where the planning, organizing, and writing of the report is guided mainly by the parts of the item or mechanism, or the steps in the process, or the features of items that caused them to be sorted into related categories. In an analytical report, however, the planning, organizing, and developing occur as the topic itself undergoes refinement and redefinition. Instead of being prescribed by the subject, the written formulation must finally come from a more abstract situation: a problem to be solved, a question to be answered, or a decision to be made. In fact, the report continually must evolve through stages, based on collected data, until it takes its final shape. And that final shape should enable readers to follow the reasoning and interpretation that led to the specific recommendations made. In short, the quality of our analysis will only be as good as the quality of the questions we ask and the answers we generate.

Even the more sophisticated student may have trouble with this notion. Close and frequent individual consultation is vital.

The *answers* students generate from their analysis can be no better than the *questions* they ask. Essential to asking the right questions is the writer's clear definition of situation, audience, and purpose. The guidelines on Master Sheet 132 should help your students.

For feasibility and problem-solving analyses as well, have each student follow the line of focus on situation, audience, and purpose.

Assign the research chapters if you have not yet done so. Be sure that bibliographies and proposals are written early, and that frequent on-the-board workshops are held on outlining tactics. Emphasize that dropping raw data into the reader's lap is not enough; evaluation and interpretation of data are crucial to the reader's understanding.

Ask for one or more progress reports (see Chapter 18, pages 350–353) during their work. Cheerleading and prodding are crucial. Set a specific and firm deadline for submitting the final draft. Require that first drafts be submitted with full attachments in case some students write an excellent report on the first round. For first submissions, require a *finished draft*, not a rough draft, to save yourself the hassle of serving as proofreader. The oral report segment at semester's end will give students time to revise as needed.

Collaborative Project 1 works well in class as a warm-up for the rigors of outlining that students will face.

Save report supplements (in Chapter 16) until students are well along in planning, researching, and writing a draft of their report. Discussion of individual supplements will be less confusing and more effective when applied to actual reports in progress.

Also, save discussions about documentation until students are well along in writing their reports. Unless you have a preference, ask them to select a documentation system.

Here are some possible topics for analysis:

- the long-term effects of a vegetarian diet
- the causes of student disinterest in campus activities
- the student transportation problem to and from your college
- comparison of two or more brands of equipment
- noise pollution from nearby airport traffic
- the effects of toxic chemical dumping in your community
- the adequacy of veterans' benefits
- the (causes, effects) of acid rain in your area
- the influence of a civic center or stadium on your community

- the best microcomputer to buy for a specific need
- the qualities employers seek in a job candidate
- the feasibility of opening a specific business
- the best location for a new business
- causes of the high dropout rate in your college
- the pros and cons of condominium ownership for you
- the feasibility of moving to a certain area of the country
- job opportunities in your field
- the effects of budget cuts on public higher education in your state
- effect of population increase on your local water supply
- the feasibility of biological pest control as an alternative to pesticides
- the feasibility of large-scale desalination of sea water as a source of fresh water
- effective water conservation measures that can be used in your area
- the effects of legalizing gambling in your state
- the causes of low morale in the company where you work part-time
- effects of thermal pollution from a local power plant on marine life
- the feasibility of converting your home to solar heating
- adequacy of police protection in your town
- the best energy-efficient, low-cost housing design for your area
- measures for improving productivity in your place of employment
- reasons for the success of a specific business in your area
- the feasibility of operating a campus food co-op
- the advisability of home birth (as opposed to hospital delivery)
- the adequacy of the evacuation plan for your area in a nuclear emergency, hurricane, or other disaster
- the feasibility and cost of improving security in the campus dorms
- the advisability of pursuing a graduate degree in your field, instead of entering the work force with a bachelor's degree.
- major causes of the wage gap between women and men in a particular field
- feasibility of door-to-door recycling pick-up in your town
- how new federal designations of wilderness areas affect the economy of your county
- feasibility of single-cell protein production as a food source
- comparative analysis of additives used in forage crop sources
- analysis of benefits from organic farming practices in improving the soil complex

- feasibility of processing specialty products from dead white pine

- advisability of using the defoliant 2,4,5-T for controlling brush in reforesting clear-cut areas

- comparative methods of soil-erosion control on irrigated farmland

- comparative analysis of careers in journalism, publishing, and technical writing

- analysis of the problems in prosecuting child abuse in your state

Master Sheets 137–157 provide the full texts and discussion of the documents that normally comprise the course project. You might discuss this material in class or place it on library reserve as assigned reading.

Master Sheet 132

Refining the Analytical Question

Situation: The primary readers have a definite problem to solve or question to answer. Whether they requested the report or you initiated the study, the reader needs to make sound decisions. Your task is to collect, evaluate, and interpret the data, to draw relevant and legitimate conclusions, and to base useful recommendations on those conclusions.

Audience: Design the report for a primary reader *other* than your writing instructor. Include a written audience-and-use profile, deciding on your level of technicality from that profile. Your writing instructor will be your secondary reader. Therefore, if the report is highly technical, adjust the supplements to a lower level of technicality for your secondary reader.

Purpose: Because your report should enable the primary reader to make a useful decision, you must address a *specific* problem or question ("Which make of heavy-duty canvas-and-wood canoe—Tromblay or Chestnut—will best serve the needs of our wilderness camp?"). You must not address a general problem or question ("How do Tromblay canoes compare with Chestnuts?"). This second type of comparison, like those found in *Consumer Reports*, is not keyed to any specific use of the equipment.

For any comparative analysis, identify at least three criteria for comparison:

- "Our canoeists are between the ages of 11 and 18. Therefore, we need canoes that can be easily carried by young people over difficult portages."
- "The terrain in the Canadian Shield region is extremely rugged. Therefore, we need highly durable canoes."
- "Many of our trips are across wide expanses of rough water. Therefore, we need the most stable and most seaworthy canoes."

Thus, the criteria for comparison, based on the needs of the primary audience (the camp's board of directors) and ranked in order of increasing *importance to the readers* are as stated here:

Tromblay and Chestnut canoes may be compared on the basis of portability, durability, and stability.

Your audience's intended use of your comparison may call for different criteria, such as *speed* and *cost* of each canoe. To serve a precise need, your study and report must be based on accurate audience-and-use analysis.

Master Sheet 133

How To Think Critically about Your Recommendations

Consider All the Details

- What exactly should be done—if anything at all?
- How exactly should it be done?
- When should it begin and be completed?
- Who will do it, and how willing are they?
- What equipment, material, or resources are needed?
- What special conditions are required?
- What will this cost, and where will the money come from?
- What consequences are possible?
- Whom do I have to persuade?

Locate the Weak Spots

- Is anything unclear or hard to follow?
- Is this course of action unrealistic?
- Is it risky or dangerous?
- Is it too complicated or costly?
- Is anything about it illegal or unethical?
- Will it cost too much?
- Will it take too long?
- Could anything go wrong?
- Who might object or be offended?

Make Improvements

- Can I rephrase anything?
- Can I change anything?
- Should I consider alternatives?
- Should I reorder my list?
- Can I overcome objections?
- Should I get advice or feedback before I submit this?

Source: Questions adapted from Ruggiero, Vincent R. *The Art of Thinking*, 5th ed. New York: Longman, 1998: 70–73.

Master Sheet 134

Chapter 25 Quiz

Name _____ Section _____

Indicate whether statements 1–5 are TRUE or FALSE by writing *T* or *F* in the blank.

1. _____ Among the typical analytical problems are "Why does X happen?" and "How do I do X?"

2. _____ An analytical report may address two or more types of analytical problems.

3. _____ The general purpose in any analysis is to prove the writer's point. In your analytical report, therefore, cite only data that support your thesis.

4. _____ You should always construct a formal outline of your analysis before writing a first draft.

5. _____ Every problem has a definite solution that can be revealed in an exhaustive analysis.

In 6–10, choose the letter of the expression that best completes each statement.

6. _____ Unlike many research reports, analytical reports (a) lead to action, (b) rely heavily on supplements, (c) always are written for clients and readers outside the organization, (d) rely only on secondary sources, or (e) always are written for readers within the organization.

7. _____ Audiences for analytical reports always expect (a) just the facts, so that they can interpret them as they see fit, (b) a full interpretation of the data, (c) valid conclusions and recommendations, (d) detailed instructions, or (e) b and c.

8. _____ Thinking critically about your recommendations means (a) to consider all the details, (b) locate the weak spots, (c) make improvements, (d) all of these, or (e) b and c.

9. _____ Audiences for analytical reports are likely to be most interested in (a) the body, (b) the conclusion, (c) the visuals and appendixes, (d) the introduction, or (e) the headings.

10. _____ Writers of analytical reports (a) begin with a clear identification of the problem or question, (b) usually discover the problem or question only *during* their investigation or drafting, (c) ignore the problem or question until they have completed their research, (d) eschew visuals, or (e) c and d.

CHAPTER
26

Oral Presentations

This brief chapter is not intended to be comprehensive in covering a subject that, in itself, would require a full semester for adequate treatment. Indeed, in a basic course (see Syllabus A) students are busy enough mastering writing skills, and might well pass over this section. Accordingly, this chapter is keyed directly to Chapter 25. It covers oral summaries of analytical reports (or formal proposals, as applicable).

After working long weeks to complete their analysis, students now have a chance to share information with others in a ten- to fifteen-minute oral report. You and your students will find the oral report one of the most enjoyable parts of the course. Speakers have the opportunity—often their first—to speak to an audience about a subject in which they have acquired solid background. Besides overcoming the usual paranoia about public speaking, the student must analyze the audience and its needs, and often must translate specialized information into a delivery that makes sense to a heterogeneous group. Moreover, because preparation time is minimal, students are left with more time to revise their proposals or analytical reports, or to struggle with other assignments that pile up at semester's end. This arrangement also gives you more time to work with individual students as they revise their written reports.

Almost invariably, students complete their summaries and brief question-and-answer sessions with a sense of elation, of having connected with the audience and shared useful information in a professional fashion. This experience is usually a far cry from that in a speech course, where the emphasis on diversity of rhetorical modes can produce superficial talks. Here, the student audience and the instructor learn things about fields often foreign to their own. In a heterogeneous class, the oral summaries form a kind of miniature crash course in the widest imaginable range of disciplines. Audience interest is high. My students attend regularly, even though I don't require attendance at this time in the semester, because they are genuinely interested in what their classmates have to say. And they are rarely disappointed. Question-and-answer sessions are lively and pertinent. In fact, you may find yourself repeatedly imposing time limits on discussions.

Require visuals. Make opaque, overhead, and slide projectors available for this purpose.

Master Sheet 135

Peer Evaluation for Oral Presentations

Oral Presentation Peer Evaluation for (name/topic) _____

COMMENTS

CONTENT

- ☐ Began with a clear purpose. _____
- ☐ Showed command of the material. _____
- ☐ Supported assertions with evidence. _____
- ☐ Used adequate and appropriate visuals. _____
- ☐ Used material suited to this audience's
  needs, knowledge, and concerns and
  interests. _____
- ☐ Acknowledged opposing views. _____
- ☐ Gave the right amount of information. _____

ORGANIZATION

- ☐ Presented a clear line of reasoning. _____
- ☐ Used transitions effectively. _____
- ☐ Avoided needless digressions. _____
- ☐ Summarized before concluding. _____
- ☐ Was clear about what the listeners should
  think or do. _____

STYLE

- ☐ Seemed confident, relaxed, and likable. _____
- ☐ Seemed in control of the situation. _____
- ☐ Showed appropriate enthusiasm. _____
- ☐ Pronounced, enunciated, and spoke well. _____
- ☐ Used appropriate gestures, tone, volume,
  and delivery rate. _____
- ☐ Had good posture and eye contact. _____
- ☐ Answered questions concisely and convincingly. _____

OVERALL PROFESSIONALISM: SUPERIOR ___   ACCEPTABLE ___   NEEDS WORK ___

Evaluator's signature: _____

Master Sheet 136

Chapter 26 Quiz

Name _____ Section _____

Indicate whether statements 1–6 are TRUE or FALSE by writing *T* or *F* in the blank.

1. _____ The impromptu delivery is usually effective for a formal report.

2. _____ Complex and elaborate visuals help keep your audience interested.

3. _____ You should use as many visuals as possible during your talk.

4. _____ For most subjects, you should distribute any handouts only *after* your presentation.

5. _____ If necessary during your talk, liven things up with clever digressions.

6. _____ Before giving your talk, you should carefully rehearse.

In 7–10, choose the letter of the expression that best completes each statement.

7. _____ Generally, the most effective way of presenting an oral report is to (a) read it aloud, (b) use extemporaneous delivery, (c) repeat it from memory, (d) give an oration, or (e) none of these.

8. _____ To keep your audience's attention, aim for a maximum delivery time of (a) 20 minutes, (b) 30 minutes, (c) 40 minutes, (d) 10 minutes, or (e) any of these.

9. _____ To be sure you stay on track during your presentation, prepare beforehand a (a) topic outline, (b) sentence outline, (c) numbered list of key words, (d) verbatim script of the entire presentation, or (e) cue cards.

10. _____ A good strategy for connecting with your audience is (a) an occasional anecdote, (b) a rich array of jokes, (c) an authoritarian tone, (d) eye contact, or (e) well-timed digressions.

Objective Test Questions

Few instructors are likely to assign every chapter in the textbook. These questions are therefore organized and labeled by chapter. Under each chapter title are several questions. For each question, students should indicate the best answer by circling the letter preceding their choice.

Chapter 1—Introduction to Technical Communication

1. A technical document focuses
   (a) mostly on the subject.
   (b) purely on the writer.
   (c) on both the subject and the writer.
   (d) on the writer's impressions.
   (e) on none of these.

2. A technical document is based on
   (a) intuition.
   (b) specialized information.
   (c) speculation.
   (d) equivocation.
   (e) all of these.

3. Errors that decrease a document's efficiency include
   (a) more (or less) information than users need.
   (b) no discernible organization.
   (c) fancier or less precise words than users need.
   (d) a and b.
   (e) all of these.

4. A technical communicator's biggest challenge is
   (a) imposing some order on the material.
   (b) selecting only that which is useful.
   (c) interpreting for users.
   (d) c and b.
   (e) all of these.

5. The *least* accurate statement about writing skills in your career is:
   (a) the higher your position in the organization, the less you will write.
   (b) good writing gives you and your ideas visibility and authority.
   (c) computers cannot give meaning to the information they transmit.
   (d) many working professionals spend at least 40 percent of their time writing or dealing with writing.
   (e) the Internet will eliminate "paper" communication by 2010.

Chapter 2—Problem Solving in Workplace Communication

1. Solving the persuasion problem means
   (a) using whatever works.
   (b) building a reasonable case.
   (c) applying diplomatic coercion.
   (d) a and d.
   (e) all of these.

2. Critical thinking
   (a) is derived from literary criticism.
   (b) is used only in emergencies.
   (c) involves weighing alternatives.
   (d) a and b.
   (e) all of these.

3. The most accurate statement below is
   (a) as long as they know the facts, people can interpret them easily.
   (b) effective communication ensures that the interests of your company take priority over the interests of your audience.
   (c) usefulness and efficiency are the ultimate measures of successful workplace communication.
   (d) documents for people outside the company usually are reviewed before they are released.
   (e) automation decreases the need for collaboration.

4. Technical communicators encounter all of the following problems except
   (a) the information problem.
   (b) the persuasion problem.
   (c) the confabulation problem.
   (d) the ethics problem.
   (e) the collaboration problem.

Chapter 3—Solving the Information Problem

1. As they approach your document, users are most interested in
   - (a) learning how smart and eloquent you are.
   - (b) finding what they need, quickly and easily.
   - (c) why the document was written.
   - (d) how well you have communicated the message.
   - (e) how well you use grammar.

2. The audience's acceptance of your document can ultimately depend on
   - (a) its level of technicality.
   - (b) the political climate in your organization.
   - (c) the individual user's attitude.
   - (d) a and c.
   - (e) all of these.

3. Writing is least likely to have informative value when it
   - (a) conveys knowledge that is new and worthwhile to the audience.
   - (b) reminds the audience of something they know but ignore.
   - (c) engages the audience's speculative faculties.
   - (d) offers fresh insight about something familiar.
   - (e) b and c.

4. The most accurate statement below is
   - (a) primary and secondary audiences often have different technical backgrounds.
   - (b) writers always should brainstorm as a very first writing step.
   - (c) primary audiences usually expect a semitechnical message.
   - (d) laypersons merely are interested in the bare facts, without explanations.
   - (e) primary audiences usually are experts.

Chapter 4—Solving the Persuasion Problem

1. An audience ideally responds to persuasion through
   - (a) compliance.
   - (b) internalization.
   - (c) obfuscation.
   - (d) elevation.
   - (e) cogitation.

2. The longest-lasting connection between persuader and audience tends to be
   - (a) the rational connection.
   - (b) the relationship connection.
   - (c) the power connection.
   - (d) the time/space connection.
   - (e) the love connection.

3. All of the following are communication constraints except
   - (a) legal constraint.
   - (b) ethical constraint.
   - (c) time constraint.
   - (d) abdominal constraint.
   - (e) social constraint.

4. Convincing evidence includes everything except
   - (a) statistics.
   - (b) examples.
   - (c) speculation.
   - (d) expert testimony.
   - (e) information that supports your claim.

5. An effective persuader avoids
   - (a) political realities.
   - (b) assorted constraints.
   - (c) extreme personas.
   - (d) anticipating audience reactions.
   - (e) conceding anything to the opponent.

Chapter 5—Solving the Ethics Problem

1. Lies that are legal in the workplace include all of the following except
   (a) promises you know you can't keep.
   (b) assurances you haven't verified.
   (c) credentials you don't have.
   (d) broken contractual promises.
   (e) inflated claims about your commitment.

2. Ethical employees always owe their greatest loyalty to
   (a) themselves.
   (b) clients and customers.
   (c) their company.
   (d) coworkers.
   (e) none of the above.

3. Ethical employees stand a better chance of speaking out and surviving if they avoid everything except
   (a) overreacting.
   (b) procrastinating.
   (c) keeping a "paper trail."
   (d) overstating the problem.
   (e) crusading.

4. Among managers polled nationwide, those who felt pressured by their company to compromise ethical standards numbered
   (a) fewer than 10 percent.
   (b) greater than 50 percent.
   (c) almost 90 percent.
   (d) 25 percent.
   (e) 2 percent.

5. Legal protection for ethical employees is
   (a) unlimited.
   (b) nonexistent.
   (c) limited.
   (d) unequivocal.
   (e) negotiable.

Chapter 6—Solving the Collaboration Problem

1. Sources of conflict in collaborative groups include
   - (a) interpersonal differences.
   - (b) gender differences.
   - (c) cultural differences.
   - (d) all of these.
   - (e) b and c.

2. Strategies for creative thinking include all of the following except
   - (a) brainstorming.
   - (b) brainwriting.
   - (c) brain scanning.
   - (d) mind mapping.
   - (e) storyboarding.

3. Electronically mediated collaboration is preferable when
   - (a) people don't know each other.
   - (b) the issue is sensitive or controversial.
   - (c) it is important to neutralize personality clashes.
   - (d) all of these.
   - (e) none of these.

4. The most accurate statement below is
   - (a) collaboration is practiced only in large companies.
   - (b) computer networks eliminate interpersonal problems in collaborative work.
   - (c) conflict in a collaborative group can be productive.
   - (d) a collaborative group functions best when each member has equal authority.
   - (e) collaboration is likely to succeed only when group members have no personal differences.

Chapter 7—Thinking Critically about the Research Process

1. The most accurate statement below is
   (a) effective research eliminates contradictory conclusions.
   (b) expert testimony usually offers a reliable "final word."
   (c) Web pages usually are dependable sources for information that offers both depth and quality.
   (d) decisions in the research process are recursive.
   (e) An ethical researcher reports all points of view as if they were equal.

2. At its deepest level, secondary research examines
   (a) trade and business publications.
   (b) the popular press.
   (c) tabloids.
   (d) specialized and government sources.
   (e) electronic newsgroups.

3. Effective research depends on all of the following except
   (a) finding a definite answer.
   (b) sampling a full range of options
   (c) getting at the facts.
   (d) achieving sufficient depth.
   (e) asking the right questions.

4. Likely sources for relatively impartial views are
   (a) trade and business sources.
   (b) Web pages.
   (c) specialized and government sources.
   (d) all of these.
   (e) none of these.

Chapter 8—Exploring Hardcopy, Online, and Internet Sources

1. Which of these is the major access tool for government publications?
   (a) *Government Reports Announcements and Index*
   (b) *The Monthly Catalog of the United States Government*
   (c) *Selected Government Publications*
   (d) *Uncle Sam's Vital Documents*
   (e) *The American Statistics Index*

2. In an automated literature search, the computer searches specific databases once the user has provided
   (a) call numbers.
   (b) key words.
   (c) ISBN numbers.
   (d) access codes.
   (e) any of these.

3. Most computerized bibliographies carry no entries before
   (a) the mid-1960s.
   (b) 1950.
   (c) 1929.
   (d) 1980.
   (e) the late-1970s.

4. Begin your research with literature that is most
   (a) specific.
   (b) technical.
   (c) general.
   (d) specialized.
   (e) readily available.

Chapter 9—Exploring Primary Sources

1. Which of these could be classified as sources for primary research?
   (a) books and articles
   (b) reports and brochures
   (c) questionnaires and interviews
   (d) any of these
   (e) none of these

2. The most accurate statement below is
   (a) to measure exactly where people stand on an issue, use closed-ended survey questions.
   (b) instead of writing out your interview questions, create a relaxed atmosphere by memorizing the questions beforehand.
   (c) use closed-ended survey questions to eliminate biased responses.
   (d) take plenty of notes during the interview.
   (e) get the most difficult, complex, or sensitive questions out of the way at the beginning of the interview.

3. The most accurate statement below is
   (a) direct observation is the surest way to eliminate bias in research.
   (b) to eliminate the potential for error, have your survey designed by a professional.
   (c) generally, the most productive way to conduct an interview is by phone.
   (d) a "sample group" should never be randomly chosen.
   (e) begin a survey with the easiest questions.

Chapter 10—Evaluating and Interpreting Information

1. The most accurate statement below is
   (a) the most recent information is almost always the most reliable.
   (b) research is most effective when it achieves "certainty."
   (c) numbers tend to be less misleading than words.
   (d) personal bias among researchers is inescapable.
   (e) "Framing" offers an ethical way to present the facts.

2. The basic criteria by which we measure the dependability of any research are
   (a) timeliness and efficiency.
   (b) validity and reliability.
   (c) conciseness and emphasis.
   (d) relevance and focus.
   (e) all of these.

3. Valid research typically produces
   (a) a conclusive answer.
   (b) a probable answer.
   (c) an inconclusive answer.
   (d) a or b.
   (e) a, b, or c.

4. Indicators of quality for a Web site include all of the following except
   (a) links to reputable sites.
   (b) material that has been peer reviewed.
   (c) options for contacting the author or organization.
   (d) presentations that include graphics, video, and sound.
   (e) objective coverage.

Chapter 11—Summarizing Information

1. The length of a summary
    (a) never should exceed 5 percent of the original.
    (b) is always specified as part of the particular writing task.
    (c) is secondary to the need for accuracy and clarity.
    (d) is none of these.
    (e) a and b.

2. When summarizing someone else's work you should
    (a) rewrite it in your own words.
    (b) document the source.
    (c) place directly quoted material within quotation marks.
    (d) do all of these.
    (e) b and c.

3. When summarizing someone else's work you should
    (a) include your personal comments as needed.
    (b) read the entire original before writing a word.
    (c) add outside material as needed.
    (d) do all of these.
    (e) b and c.

4. Significant material in a summary includes
    (a) conclusions and recommendations.
    (b) visuals.
    (c) background discussions.
    (d) all of these
    (e) b and c.

5. Summaries sometimes are called
    (a) descriptive abstracts.
    (b) informative abstracts.
    (c) none of these.
    (d) all of these.
    (e) ancillary correspondence.

6. Above all, a good summary is accurate and
    (a) brief.
    (b) grammatical.
    (c) engaging.
    (d) none of these.
    (e) concise.

7. For a summary, the essential message does not include
    (a) controlling ideas.
    (b) conclusions and recommendations.
    (c) examples and visuals.
    (d) any of these.
    (e) b and c.

8. For a large and unspecified audience, to avoid confusion you should make your summary
   (a) highly technical.
   (b) nontechnical.
   (c) as brief as possible.
   (d) none of these.
   (e) a and c.

9. A descriptive abstract reflects
   (a) what the original contains.
   (b) what the original is about.
   (c) the essential message from the original.
   (d) all of these.
   (e) none of these.

Chapter 12—Organizing for Users

1. Successful communicators generally spend more time
   - (a) writing than thinking.
   - (b) relaxing than planning.
   - (c) thinking than writing.
   - (d) net surfing than brainstorming.
   - (e) drafting than revising.

2. For notation in a formal outline, use
   - (a) the decimal system.
   - (b) the alphanumeric system.
   - (c) the monosyllabic typology system.
   - (d) the obtuse labeling system.
   - (e) a or b.

3. The sequence you choose for your outline should depend on
   - (a) what you feel is most creative.
   - (b) what your users will find imaginative.
   - (c) the order in which you expect users to approach the material.
   - (d) the order your readers will find most suspenseful.
   - (e) a and b.

4. In writing the introduction to your report,
   - (a) make it as long as you can.
   - (b) keep it as brief as you can.
   - (c) always include a full list of works cited.
   - (d) give users as much background as you have.
   - (e) never mention any limitations your report might have.

5. For a brief document, use
   - (a) a formal topic outline.
   - (b) a decimal outline.
   - (c) an informal outline.
   - (d) a predicated outline.
   - (e) no outline.

6. Beyond expecting worthwhile content, users expect a message to be
   - (a) impressive.
   - (b) well adorned.
   - (c) accessible.
   - (d) all of these.
   - (e) none of these.

7. In a unified paragraph,
   - (a) everything expands the main point.
   - (b) the thought sequence is clearly connected.
   - (c) the subject is appropriate to the user's needs.
   - (d) the conclusion is inventive.
   - (e) the conclusion is persuasive.

8. The shape of an organized unit of meaning is best illustrated in the form of
    (a)   a transitional paragraph.
    (b)   a complaint letter.
    (c)   a standard paragraph.
    (d)   a long paragraph.
    (e)   an internal memo.

9. Most paragraphs in professional writing
    (a)   end with a topic sentence.
    (b)   begin with a topic sentence.
    (c)   require no topic sentence.
    (d)   have two or more topic sentences.
    (e)   have three or more topic sentences.

10. Topic sentences
    (a)   should generally be no longer than ten words.
    (b)   are known also as orienting sentences.
    (c)   should only hint at the main point.
    (d)   are all of these.
    (e)   are none of these.

11. To attract attention to an important idea,
    (a)   use a long paragraph.
    (b)   use a series of long paragraphs.
    (c)   use a short paragraph.
    (d)   underline the entire paragraph.
    (e)   do any of these.

12. Ways of damaging coherence include
    (a)   using too many short, choppy sentences.
    (b)   placing sentences in the wrong order.
    (c)   using insufficient transitions.
    (d)   a and b.
    (e)   a, b, and c.

13. A spatial order of paragraph development is most useful for
    (a)   describing an event.
    (b)   concluding a problem-solving analysis.
    (c)   describing an item or a mechanism.
    (d)   giving instructions.
    (e)   none of these.

14. A standard paragraph generally follows this organizing pattern:
    (a)   introduction, body, conclusion.
    (b)   topic sentence and abstract.
    (c)   introduction and extended example.
    (d)   argument, refutation, and support.
    (e)   none of these.

15. A topic sentence
    (a)   provides details that explain the main point.
    (b)   provides an orienting framework.
    (c)   concludes the argument.
    (d)   previews subsequent paragraphs.
    (e)   establishes counterpoint.

Chapter 13—Revising for Readable Style

1. In a technical document, avoid
   (a) telegraphic writing.
   (b) the active voice.
   (c) direct address.
   (d) conciseness.
   (e) all of these.

2. Prefer the passive voice
   (a) to report bad news.
   (b) when the actor is unknown.
   (c) in giving instructions.
   (d) in a job application.
   (e) all of these.

3. Achieve conciseness by
   (a) eliminating negative constructions.
   (b) substituting nouns for verbs.
   (c) replacing words with phrases.
   (d) using passive constructions.
   (e) all of these.

4. To combine a series of short, choppy sentences, use
   (a) coordination.
   (b) subordination.
   (c) either a or b.
   (d) association.
   (e) annunciation.

5. Redundancy, needless repetition, and clutter words primarily harm a sentence's
   (a) fluency.
   (b) clarity.
   (c) rhythm.
   (d) conciseness.
   (e) grammaticality.

6. In its style features, an efficient sentence is clear, concise, and
   (a) entertaining.
   (b) informative.
   (c) fluent.
   (d) prosaic.
   (e) harmonious.

7. For best emphasis, avoid placing the key word or phrase at the sentence's
   (a) beginning.
   (b) middle.
   (c) end.
   (d) terminal position.
   (e) all of these.

8. When combining sentences, place the idea that deserves the most emphasis in a clause that is
   (a) dependent.
   (b) subordinate.
   (c) independent.
   (d) indirect.
   (e) direct.

9. Generally you should avoid
   (a) analogies.
   (b) euphemisms.
   (c) the active voice.
   (d) fluent expression.
   (e) direct address.

10. For most workplace documents, choose a tone that is
    (a) formal.
    (b) conversational.
    (c) serious.
    (d) prosaic.
    (d) inspirational.

Chapter 14—Designing Visuals

1. Well-chosen visuals cause the presenter to seem
   (a) prepared.
   (b) credible.
   (c) persuasive.
   (d) a and b.
   (e) a, b, and c.

2. Which of these is not a benefit of using visuals in a professional document?
   (a) increased reader interest.
   (b) condensed information.
   (c) entertainment value.
   (d) emphasis.
   (e) clarity.

3. Any visual should
   (a) repeat no information already given in the text.
   (b) stand independently.
   (c) occupy no more than half a page.
   (d) precede its discussion.
   (e) entertain.

4. An electronically enhanced visual should
   (a) stimulate the audience with abundant textures and patterns.
   (b) exhibit restraint and good taste.
   (c) demonstrate the writer's computer expertise.
   (d) do all of these.
   (e) do none of these.

5. Computer graphics allow for
   (a) experimentation.
   (b) refinement.
   (c) design options.
   (d) all of these.
   (e) b and c.

6. For illustrating a trend, the appropriate figure is typically a
   (a) line graph.
   (b) bar graph.
   (c) pie chart.
   (d) column chart.
   (e) flowchart.

7. All visuals belong in
   (a) the report text, as close as possible to their discussion.
   (b) appendixes.
   (c) in either a or b, depending on their relationship to the discussion.
   (d) the glossary, after the definitions.
   (e) the report's introduction.

8. A legend is a
   (a) caption that explains each bar or line in a graph.
   (b) prose introduction to a visual.
   (c) list that credits data sources for the visual.
   (d) visual of historic value.
   (e) none of these.

9. To show how the parts of an item are assembled, use
   (a) an organization chart.
   (b) a photograph.
   (c) an exploded diagram.
   (d) a pictogram.
   (e) a pie chart.

10. An outstanding benefit of computer graphics is
    (a) higher employee motivation.
    (b) the endless capacity for "what if" projections.
    (c) the many colors that can be produced on a map, graph, and so on.
    (d) composing ease.
    (e) clip art.

Chapter 15—Designing Pages and Documents

1. The inaccurate statement here is
   (a) technical information usually is designed differently from other forms of writing.
   (b) technical documents rarely get the audience's undivided attention.
   (c) most people look forward to reading work-related documents.
   (d) in the computer age, any document is forced to compete for the audience's attention.
   (e) none of these.

2. Computer technology
   (a) has decreased the writer's responsibility in the workplace.
   (b) has increased the need for secretaries.
   (c) has heightened audience expectations about document formats.
   (d) has all but eliminated mechanical errors from documents.
   (e) all of these.

3. White space in a document should be
   (a) random.
   (b) impressionistic.
   (c) deliberately designed.
   (d) eliminated.
   (e) formless.

4. "Justified text" means
   (a) uneven right margins.
   (b) elevated gutter space.
   (c) a persuasive margin.
   (d) even left and right margins.
   (e) ragged right margin.

5. To set off listed items, use
   (a) numbers, dashes, or bullets.
   (b) full caps.
   (c) ornate typefaces.
   (d) visionary punctuation.
   (e) any of these.

Chapter 16—Adding Document Supplements

1. A letter of transmittal should do all of these except
   - (a) indicate pride and satisfaction in the writer's work.
   - (b) apologize for possible inadequacies in the document.
   - (c) acknowledge those who helped with the document.
   - (d) refer readers to sections of special interest.
   - (e) express the writer's willingness to answer any questions.

2. In your table of contents, you should do all these except
   - (a) list the title page.
   - (b) phrase headings identically to those in the report text.
   - (c) use spaced horizontal dots to connect headings to page numbers.
   - (d) list end matter.
   - (e) include no headings that are not used as headings in the report text.

3. A supplement that precedes the report is the
   - (a) glossary.
   - (b) white space.
   - (c) abstract.
   - (d) works-cited page.
   - (e) a, c, and d.

4. The writer's "off-the-record" comments belong in the
   - (a) informative abstract.
   - (b) letter of transmittal.
   - (c) appendix.
   - (d) glossary.
   - (e) none of these.

5. Items that do not belong in an appendix include
   - (a) interview questions and responses.
   - (b) recommendations.
   - (c) formulas and calculations.
   - (d) maps.
   - (e) all of these.

Chapter 17—Testing the Usability of Your Document

1. The most accurate statement below is
   - (a) usability testing is usually performed only on complex products or documents.
   - (b) online documents often are harder to navigate than printed pages.
   - (c) human factors tend to enhance performance.
   - (d) expert users tend to read a document sequentially (page by page).
   - (e) a basic usability survey focuses primarily on a document's content.

2. Usability criteria for a document are defined by
   - (a) the type of task.
   - (b) user characteristics.
   - (c) constraints of the setting.
   - (d) none of these.
   - (e) all of these.

3. A usable document enables users to do all of the following *except*
   - (a) easily locate the needed information.
   - (b) understand the information.
   - (c) use the information successfully.
   - (d) develop creative approaches to the task.
   - (e) carry out the task safely, efficiently, and accurately.

Chapter 18—Memo Reports and Electronic Mail

1. The most accurate statement below is
   - (a) as a form of internal correspondence, memos have few legal implications.
   - (b) email increases both the quantity and quality of information.
   - (c) monitoring of email by an employer is legal.
   - (d) writing in FULL CAPS increases the readability of email.
   - (e) a memorandum should be no more than one page long.

2. Memos are major means of written communication within organizations because they
   - (a) leave a "paper trail."
   - (b) are easy to write and read.
   - (c) are less expensive than other communications media.
   - (d) a and c.
   - (e) b and c.

3. A typical memo does not have
   - (a) a complimentary close and signature.
   - (b) a subject line.
   - (c) topic headings.
   - (d) a distribution notation.
   - (e) single spacing.

4. Use your company's email network to send
   - (a) a formal letter to a client.
   - (b) an evaluation of an employee.
   - (c) an announcement for a company picnic.
   - (d) a dinner invitation to a colleague.
   - (e) any of these.

Chapter 19—Letters and Employment Correspondence

1. When providing employment references, never
    (a) list the names of references on your resume.
    (b) use an acquaintance as a reference.
    (c) list more than two references.
    (d) list a person whose permission you have not yet received.
    (e) do b, c, and d.

2. The standard parts of a letter include everything but
    (a) inside address.
    (b) heading.
    (c) salutation.
    (d) enclosure notation.
    (e) complimentary close.

3. Letterese
    (a) is an engaging feature of letter style.
    (b) is a way to connect with a reader.
    (c) makes letters seem unimaginative and boring.
    (d) makes readers feel comfortable.
    (e) is the clearest way to communicate your point.

4. When you want the recipient to know immediately the point of your letter, use
    (a) opening conventions.
    (b) the direct plan.
    (c) the indirect plan.
    (d) the forward plan.
    (e) the assertive plan.

5. In screening candidates' résumés, generally employers
    (a) study them for mechanical errors.
    (b) look for elegant paper and typeface.
    (c) spend less than 60 seconds.
    (d) look for imaginative formatting.
    (e) do b, c, and d.

Chapter 20—Web Pages and Other Electronic Documents

1. The most accurate statement below is
   (a) users expect Web documents to offer a traditional introduction, discussion, and conclusion—in the same way a paper document is organized.
   (b) for effective Web page design, the more graphs and special effects, the better.
   (c) online text should ordinarily be at least 50 percent shorter than its hard copy equivalent.
   (d) persuasive promotional writing increases a Web site's usability.
   (e) Web site users prefer as many navigational choices as possible.

2. Documents on the World Wide Web offer all the following advantages *except*
   (a) interactivity.
   (b) reciprocal use.
   (c) porousness.
   (d) continuous and rapid updating.
   (e) more rapid reading time than for paper documents.

3. The name for the interactive system that allows users to explore a topic from countless angles and to various depths is
   (a) online documentation.
   (b) electronic mail.
   (c) hypertext.
   (d) fiber optics.
   (e) electromagnetic stimulation.

4. Instructions in computer-based training are provided by
   (a) FAX network.
   (b) online documentation.
   (c) telecommuting.
   (d) boilerplate.
   (e) browsers.

5. Types of online documentation include
   (a) error messages.
   (b) reference guides.
   (c) tutorial lessons.
   (d) help and review options.
   (e) all of these.

Chapter 21—Technical Definitions

1. To provide a *general* definition of a specialized term, use
   (a) a sentence definition.
   (b) an expanded definition.
   (c) a parenthetical definition.
   (d) a or c.
   (e) a or b.

2. In your report, always place expanded definitions
   (a) in the introduction.
   (b) in the report body.
   (c) in the appendix.
   (d) in the conclusion.
   (e) in none of these.

3. When expanding a definition, use
   (a) every expansion strategy described in Chapter 21.
   (b) as many expansion strategies as possible.
   (c) only the expansion strategies that serve the user's needs.
   (d) no more expansion strategies than you have space to accommodate.

4. Working definitions often are expressed as
   (a) parenthetical definitions.
   (b) imperative statements.
   (c) sentence definitions.
   (d) expanded definitions.
   (e) subjective definitions.

5. A useful definition in technical writing is
   (a) objective.
   (b) pontifical.
   (c) written in plain English.
   (d) creative.
   (e) a and c.

Chapter 22—Technical Descriptions and Specifications

1. Users of technical descriptions usually have all these questions except
    (a) What does it do?
    (b) What is your opinion of the item?
    (c) What does it look like?
    (d) How does it work?
    (e) What is it made of?

2. When you cannot identify your audience, you should
    (a) use details at the lowest level of technicality.
    (b) remember that too many details are better than too few.
    (c) strive for subjective impartiality.
    (d) do a and b.
    (e) do a, b, and c.

3. Which of these is not a descriptive sequence?
    (a) spatial
    (b) chronological
    (c) functional
    (d) metaphysical

4. The most precise technical descriptions are
    (a) intuitive.
    (b) impartial.
    (c) allegorical.
    (d) neoscientific.
    (e) vividly subjective.

5. To use "hand-propelled graphite transcription device" instead of "lead pencil" as part of your descriptive terminology is
    (a) more precise.
    (b) more elegant.
    (c) more impressive.
    (d) needlessly complicated.
    (e) rigorously semantical.

Chapter 23—Procedures and Processes

1. The less specialized the users of your instructions are,
   (a) the fewer visuals they need.
   (b) the more visuals they need.
   (c) the fewer steps they need.
   (d) the more steps they need.
   (e) the fewer examples they need.

2. Which of these guidelines is wrong?
   (a) Give users no more than they need to complete a step.
   (b) Adjust the information rate to your audience's understanding.
   (c) Provide visuals to reinforce the prose.
   (d) Omit steps users will find obvious.
   (e) Strive above all for brevity.

3. The only items that should interrupt the steps in a set of instructions are
   (a) warnings.
   (b) cautions.
   (c) notes.
   (d) definitions.
   (e) all of these.
   (f) a, b, and c.

4. Phrase all instructions in
   (a) the active voice and imperative mood.
   (b) the passive voice and indicative mood.
   (c) the active voice and indicative mood.
   (d) the active voice and subjunctive mood.
   (e) the passive voice and submissive mood.

5. To ensure an accessible format for your instructions, avoid
   (a) informative headings.
   (b) numbered lists.
   (c) highlighting.
   (d) white space.
   (e) none of these.

Chapter 24—Proposals

1. Proposals can be classified according to
   (a) origin.
   (b) audience.
   (c) intention.
   (d) all of these.
   (e) only a and b.

2. The proposal abstract should contain
   (a) a statement of the problem and causes.
   (b) proposed solutions.
   (c) an assessment of the plan's feasibility.
   (d) all of these.
   (e) only a and b.

3. If the primary audience for your proposal is expert or informed and the secondary audience has no expertise,
   (a) write the proposal itself for a lay audience, and provide appendixes with technical details.
   (b) keep the proposal itself technical, and provide supplements with less technical details for secondary readers.
   (c) develop two versions of your proposal, one technical and the other nontechnical.
   (d) write the entire document (including supplements) at the lowest level of technicality.
   (e) do none of these.

4. The proposal introduction typically has these subsections:
   (a) timetable and request for action.
   (b) materials, equipment, and cost.
   (c) statement of problem and need.
   (d) all of these.
   (e) none of these.

5. Clients expect proposals to
   (a) be at least five pages long.
   (b) have many attractive visuals.
   (c) be written at the lowest level of technicality.
   (d) all of these.
   (e) none of these.

6. Guidelines for developing a specific proposal are included in
   (a) the RFP.
   (b) the proposing firm's operations manual.
   (c) any comprehensive textbook.
   (d) the contract bylaws.
   (e) all of these.

7. A proposal for improving your company's employee morale, requested by the vice-president in charge of personnel, would probably be classified as
   (a) a solicited internal research proposal.
   (b) an unsolicited external planning proposal.
   (c) a solicited internal planning proposal.
   (d) a modest proposal.
   (e) none of these.

8. Especially important to nonspecialized reviewers of a long proposal is (are) the
   (a) conclusion.
   (b) appendixes.
   (c) visuals.
   (d) abstract.
   (e) none of these.

9. Successful proposals are usually those that are most
   (a) specific.
   (b) ambitious.
   (c) elaborately designed.
   (d) visually oriented.
   (e) all of these.

10. Besides being clear, the proposal plan must be
   (a) inexpensive.
   (b) inspirational.
   (c) optimistic.
   (d) technically unfathomable.
   (e) none of these.

Chapter 25—Analytical Reports

1. For your report, select only the sources that are
   (a) reputable.
   (b) reasonably impartial.
   (c) authoritative.
   (d) all of these.
   (e) none of these.

2. A conclusion to an analytical report
   (a) never appeals to the audience's emotions.
   (b) may appeal secondarily to emotions, but always appeals primarily to reason.
   (c) should be noncommittal.
   (d) is all of these.
   (e) is none of these.

3. During your analysis, you must ask yourself these question(s) about your data:
   (a) Is this reliable and important?
   (b) What does it mean?
   (c) What action is needed?
   (d) Who should take action?
   (e) all of these.

4. A feasibility analysis is designed primarily to
   (a) solve a problem.
   (b) compare two or more similar items.
   (c) assess the practicality of an idea or plan.
   (d) sell a product.
   (e) do none of these.

5. Audiences for analytical reports expect
   (a) just the facts, so that they can interpret them as they see fit.
   (b) full interpretation of the data.
   (c) definite conclusions and recommendations.
   (d) b and c.
   (e) a and b.

6. Audiences for analytical reports are likely to be most interested in
   (a) the body.
   (b) conclusions and recommendations.
   (c) visuals and appendixes.
   (d) headings.
   (e) none of these.

Chapter 26—Oral Presentations

1. For an impromptu speech, you
   (a) read from prepared notes.
   (b) rely on memorized text.
   (c) do not prepare.
   (d) speak persuasively.
   (e) use an outline.

2. Extemporaneous speaking allows the speaker to
   (a) speak conversationally.
   (b) memorize the delivery.
   (c) omit visuals.
   (d) read from a carefully prepared text.
   (e) omit an outline.

3. To use visuals effectively in oral reports you should
   (a) keep them simple.
   (b) allow the audience to interpret the visuals on their own.
   (c) distribute copies of all visuals before your presentation.
   (d) make the visuals as elaborate as possible.
   (e) use as many as possible.

4. When preparing an extemporaneous report, you should
   (a) memorize your material.
   (b) identify your audience.
   (c) rehearse often.
   (d) do all of these.
   (e) do b and c.

5. Oral reports have all but which of these advantages?
   (a) They require no planning.
   (b) They elicit immediate feedback.
   (c) They can be more engaging than written reports.
   (d) all of these.
   (e) none of these.

Answers to
Objective Test Questions

Chapter 1
1. a
2. b
3. e
4. e
5. a

Chapter 2
1. b
2. c
3. d
4. c

Chapter 3
1. b
2. e
3. c
4. a

Chapter 4
1. b
2. a
3. d
4. c
5. c

Chapter 5
1. d
2. e
3. c
4. b
5. c

Chapter 6
1. d
2. c
3. d
4. c

Chapter 7
1. d
2. d
3. a
4. c

Chapter 8
1. b
2. b
3. a
4. a

Chapter 9
1. c
2. a
3. e

Chapter 10
1. d
2. b
3. e
4. d

Chapter 11
1. c
2. d
3. b
4. a
5. b
6. e
7. c
8. b
9. b

Chapter 12
1. c
2. e
3. c
4. b
5. c
6. c
7. a
8. c
9. b
10. b
11. c
12. e
13. c
14. a
15. b

Chapter 13

1. a
2. b
3. a
4. c
5. d
6. c
7. b
8. c
9. b
10. b

Chapter 14

1. e
2. c
3. b
4. b
5. a
6. a
7. c
8. a
9. c
10. b

Chapter 15

1. c
2. c
3. c
4. d
5. a

Chapter 16

1. b
2. a
3. c
4. b
5. b

Chapter 17

1. b
2. e
3. d

Chapter 18

1. c
2. a
3. a
4. c

Chapter 19

1. d
2. d
3. c
4. b
5. c

Chapter 20

1. c
2. e
3. c
4. b
5. e

Chapter 21

1. d
2. e
3. c
4. c
5. e

Chapter 22

1. b
2. d
3. d
4. b

Chapter 23

1. b
2. e
3. f
4. a
5. e

Chapter 24

1. d
2. d
3. b
4. c
5. e
6. a
7. c
8. d
9. a
10. e

Chapter 25

1. d
2. b
3. e
4. c
5. d
6. b

Chapter 26

1. c
2. a
3. a
4. e
5. a

Answers to
Chapter Quiz Questions

Ch. 1 Quiz (Master Sheet 17)
1. F
2. F
3. F
4. T
5. F
6. meaning
7. communication
8. a
9. b
10. b

Ch. 2 Quiz (Master Sheet 20)
1. F
2. F
3. T
4. T
5. T
6. F
7. c
8. e
9. b
10. c

Ch. 3 Quiz (Master Sheet 28)
1. T
2. T
3. F
4. T
5. F
6. F
7. Why do they want the report?
   How much do they already know?
   What is the technical background of the primary and secondary audience?
   When is the report due?
   What does the audience need to know?
8. In college, the audience you write for knows more than you, and is testing your knowledge; on the job, the audience you write for knows less, and is using your knowledge.
9. Writing has informative value when it conveys knowledge that is new and worthwhile, or when it has fresh insight about something familiar.
10. Primary audiences usually are those who requested the document and who will use it as a basis for decisions or action. Secondary audiences are those who will carry out the project, who will advise the primary readers about their decision, or who will somehow be affected by this decision.

Ch. 4 Quiz (Master Sheet 32)
1. F
2. F
3. T
4. T
5. F
6. F
7. b
8. a
9. d
10. c

Ch. 5 Quiz (Master Sheet 39)
1. F
2. F
3. F
4. F
5. T
6. F
7. d
8. e
9. c
10. b

Ch 6. Quiz (Master Sheet 42)
1. F
2. T
3. F
4. T
5. T
6. F
7. F
8. d
9. c
10. d

Ch. 7 Quiz (Master Sheet 47)
1. F
2. F
3. F
4. T
5. T
6. F
7. F
8. d
9. a
10. c

Ch. 8 Quiz (Master Sheet 48)
1. T
2. F
3. T
4. F
5. T
6. T
7. T
8. T
9. a
10. a

Ch. 9 Quiz (Master Sheet 49)
1. T
2. F
3. F
4. F
5. F
6. T
7. F
8. F
9. F
10. F

Ch. 10 Quiz (Master Sheet 50)
1. F
2. F
3. T
4. F
5. T
6. F
7. F
8. b
9. e
10. d

Ch. 11 Quiz (Master Sheet 53)
1. F
2. F
3. F
4. T

5. F
6. T
7. c
8. c
9. b
10. a

Ch. 12 Quiz (Master Sheet 57)
1. F
2. F
3. T
4. F
5. T
6. T
7. a
8. c
9. b
10. c

Ch. 13 Quiz (Master Sheet 82)
1. F
2. F
3. F
4. F
5. T
6. c
7. b
8. c
9. b
10. b

Ch. 14 Quiz (Master Sheet 94)
1. F
2. F
3. T
4. T
5. F
6. a
7. c
8. a
9. e
10. b

Ch. 15 Quiz (Master Sheet 98)
1. T
2. T
3. F
4. F
5. F
6. c
7. b
8. b

9. a
10. b

Ch. 16 Quiz (Master Sheet 99)
1. F
2. F
3. F
4. T
5. T
6. d
7. b
8. b
9. c
10. b

Ch. 17 Quiz (Master Sheet 100)
1. F
2. T
3. T
4. T
5. T
6. F
7. F
8. e
9. d
10. attitudes, abilities, limitations

Ch. 18 Quiz (Master Sheet 101)
1. F
2. F
3. T
4. F
5. F
6. F
7. a
8. a
9. b
10. c

**Ch. 19 Quiz A
(Master Sheet 118)**
1. c
2. c
3. a
4. b
5. d
6. c
7. b
8. F
9. F
10. T

**Ch. 19 Quiz B
(Master Sheet 119)**
1. b
2. b
3. b
4. e
5. c
6. b
7. b
8. a
9. b
10. b

Ch. 20 Quiz (Master Sheet 120)
1. F
2. F
3. T
4. T
5. F
6. T
7. e
8. c
9. b
10. e

Ch. 21 Quiz (Master Sheet 121)
1. T
2. F
3. F
4. c
5. e
6. b
7. b
8. b
9. term, class, distinguishing features
10. negation, examples, visuals, analysis, comparison or contrast, operating principle, special conditions

Ch. 22 Quiz (Master Sheet 123)
1. F
2. F
3. F
4. F
5. T
6. T
7. d
8. c
9. What does it do?
What does it look like?

What is it made of?
How does it work?
10. Functional and chronological

Ch. 23 Quiz (Master Sheet 125)
1. F
2. F
3. F
4. T
5. b
6. b
7. a
8. a
9. c
10. c

Ch. 24 Quiz (Master Sheet 131)
1. T
2. F
3. T
4. T
5. F
6. a
7. c
8. d
9. a
10. e

Ch. 25 Quiz (Master Sheet 134)
1. F
2. T
3. F
4. F
5. F
6. a
7. b
8. d
9. b
10. a

Ch. 26 Quiz (Master Sheet 136)
1. F
2. F
3. F
4. T
5. F
6. T
7. b
8. a
9. b
10. d

Master Sheet 137

A Sample Research Project: A Document Sequence Culminating in the Final Report

Professionals in the workplace engage in one common and continual activity; they struggle with big decisions like these:

- Should we promote Jones to general manager, or should we bring in someone from outside the company?
- Why are we losing customers, and what can we do about it?
- How can we decrease work-related injuries?
- Should we encourage foreign investment in our corporation?
- Will this employee-monitoring program ultimately increase productivity or alienate our workers?

Research projects routinely are undertaken to answer questions like those above—to provide decision makers the information they need.

A course research project, like any in the workplace, is designed to fill a specific information need: to answer a question, to solve a problem, to recommend a course of action. And so students of technical communication often spend much of the semester preparing a research project.

Virtually any major project in the workplace requires that all vital information at various stages (the plan, the action taken, the results) be recorded *in writing*. The proposal/progress report/final report sequence keeps the audience informed at each stage of a project.

Master Sheet 138

A Sample Research Project[1]

Mike Cabral, communications major, works part-time as assistant to the production manager for *Megacrunch*, a computer magazine specializing in small-business applications. (*Production* is the transforming of manuscripts into a published form.) Mike's writing instructor assigns a course project and encourages students to select topics from the workplace, when possible. Mike asks *Megacrunch* production manager Marcia White to suggest a research topic that might be useful to the magazine.

White outlines a problem she thinks needs careful attention: Now six years old, *Megacrunch* has enjoyed steady growth in sales volume and advertising revenue—until recently. In the past year alone, *Megacrunch* has lost $150,000 in subscriptions and one advertising account worth $60,000. White knows that most of these losses are caused by increasing competition (three competing magazines have emerged in 18 months). In response to these pressures, White and the executive staff have been exploring ways of reinvigorating the magazine—through added coverage and "hot" features, more appealing layouts and page design, and creative marketing. But White is concerned about another problem that seems partially responsible for the fall in revenues: too many errors are appearing in recent issues.

When *Megacrunch* first hit the shelves, the occasional error in grammar or accuracy seemed unimportant. But as the magazine grew in volume and complexity, the errors increased. Errors in recent issues include misspellings, inaccurate technical details, unintelligible sentences and paragraphs, and scrambled source code (in sample programs).

White asks Mike if he's interested in researching the error problem and looking into quality-control measures. She feels that his three years' experience with the magazine qualifies him for the task. Mike accepts the assignment.

White cautions Mike that this topic is politically sensitive, especially to the editorial staff and to the investors. She wants to be sure that Mike's investigation doesn't merely turn up a lot of dirty laundry. Above all, White wants to preserve the confidence of investors—not to mention the morale of the editorial staff, who do a good job in a tough environment, plagued by impossible deadlines and constant pressure. White knows that offending people—even unintentionally—can be disastrous.

She therefore insists that all project documents express a supportive rather than critical point of view: "What could we be doing better?" instead of "What are we doing wrong?" Before agreeing to release the information from company files (complaint letters, notes from irate phone calls, and so on) White asks Mike to submit a proposal, in which the intent of this project is made absolutely clear.

---

[1] In the interest of privacy, the names of all people, publications, companies, and products in Mike's documents have been changed.

Master Sheet 139

The Project Documents

Three types of documents lend shape and sequence to the research project: **the proposal,** which spells out the plan; **the progress report,** which keeps track of the investigation; and **the final report,** which analyzes the findings. This presentation shows how these documents function together in Mike Cabral's reporting process.

The Proposal Stage

Proposals offer plans for meeting needs. A proposal's primary audience consists of those who will decide whether to approve, fund, or otherwise support the project. Reviewers of a research proposal usually begin with questions like these:

- What, exactly, do you intend to find out?
- Why is the question worth answering, or the problem worth solving?
- What benefits can we expect from this project?

Once reviewers agree the project is worthwhile, they will want to know all about the plan:

- How, exactly, do you plan to do it?
- Is the plan realistic?
- Is the plan acceptable?

Besides these questions, reviewers may have others: *How much will it cost? How long will it take? What makes you qualified to do it?* and so on. See Chapter 24 for more discussion and examples.

Mike Cabral knows his proposal will have only Marcia White (and possibly some executive board members) as the primary audience. The secondary audience is Mike's writing instructor, who must approve the topic as well. And at some point, Mike's documents could find their way to his co-workers.

White already knows the background and she needs no persuading that the project is worthwhile, but she does expect a realistic plan before she will approve the project. (For his instructor, Mike attaches a short appendix [not shown here] outlining the background and his qualifications.) Also, Mike concentrates on his emphasis: he wants the proposal to be positive rather than critical, so as not to offend anyone. He therefore focuses on achieving *greater accuracy* rather than *fewer errors.* So that his instructor can approve the project, Mike submits the proposal by the semester's fourth week.

Master Sheet 140

TO: Marcia White, Production Manager   September 26, 19xx

FROM: Mike Cabral, Production Assistant *MC*

SUBJECT: Proposal for Studying Ways to Improve Quality Control at *Megacrunch*

Introduction

Tells what the problem is, and what it means to the company

The growing number of grammatical, informational, and technical errors in each monthly issue of *Megacrunch* is raising complaints from authors, advertisers, and readers. Beyond compromising the magazine's reputation for accurate and dependable information, these errors—almost all of which seem avoidable—endanger our subscription and advertising revenues.

Summary of the Problem

Further defines the problem and its effects

Authors are complaining of errors in published versions of their articles. Software developers assert that errors in reviews and misinformation about products have damaged reputations and sales. For example, Osco Scientific, Inc. claims to have lost $150,000 in software sales because of an erroneous review in *Megacrunch*.

Although we continue to receive a good deal of "fan mail," we also receive letters speculating about whether *Megacrunch* has lost its edge as a leading resource for small-business users.

Proposed Study

Describes the project and tells who will carry it out

I propose to examine the errors that most frequently recur in our publication, to analyze their causes, and to search for ways of improving quality.

Methods and Sources

Describes the research strategy

In addition to close examination of recent *Megacrunch* issues and competing magazines, my primary data sources will include correspondence and other feedback now on file from authors, developers, and readers. I also plan telephone interviews with some of the above sources. In addition, interviews with our editorial staff should yield valuable insights and suggestions. As secondary material, books and articles on editing and writing can provide sources of theory and technique.

Conclusion

Encourages audience support by telling why the project will be beneficial

We should not allow avoidable errors to eclipse the hard work that has made *Megacrunch* the leading Cosmo resource for small-business users. I hope my research will help resolve many such errors. With your approval, I will begin immediately.

Master Sheet 141

The Progress-Report Stage

The progress report keeps the audience up-to-date on the project's activities, new developments, accomplishments or setbacks, and timetable. Depending on the size and length of the particular project, the number of progress reports will vary. (Mike's course project will require only one.) The audience approaches any progress report with two big questions:

- Is the project moving ahead according to plan and schedule?
- If not, why not?

The audience may have various subordinate questions as well. See pages 350–353 for more discussion and examples.

Mike Cabral designs his progress report for his boss *and* his instructor, and turns it in by the semester's tenth week. Each of these users will want to know what Mike has accomplished so far.

Master Sheet 142

Rangeley Publishing Company

TO: Marcia White, Production Manager November 6, 19xx

FROM: Mike Cabral, Production Assistant *MC*

SUBJECT: <u>Report of Progress on My Research Project: A Study of Ways for Improving Quality Control at *Megacrunch* Magazine</u>

Work Completed

My topic was approved on September 28, and I immediately began both primary and secondary research. I have since reviewed file letters from contributors and readers, along with notes from phone conversations with various clients and from interviews with *Megacrunch's* editing staff. I have also surveyed the types and frequency of errors in recent issues. Recent books and articles on writing and editing have rounded out my study. The project has moved ahead without complications. With my research virtually completed, I have begun to interpret the findings.

Preliminary Interpretation of Findings

From my primary research and my own editing experience, I am developing a focused idea of where some of the most avoidable problems lie and how they might be solved. My secondary sources offer support for the solutions I expect to recommend, and they suggest further ideas for implementing the recommendations. With a realistic and efficient plan, I think we can go a long way toward improving our accuracy.

Work Remaining

So far, the project is on schedule. I plan to complete the interpretation of all findings by the week of November 29, and then to organize, draft, and revise my final report in time for the December 14 submission deadline.

(Margin notes:)

Tells what has been done so far, and what is now being done

Tells what has been found so far, and what it seems to mean

Assesses the project schedule; describes work remaining and gives a completion date

Master Sheet 143

The Final-Report Stage

The final report presents the results of the research project: findings, interpretations, and recommendations. This document answers questions like these:

- What did you find?
- What does it all mean?
- What should we do?

Depending on the topic and situation, of course, the audience will have specific questions as well. See Chapter 25 for discussion and examples.

During his research, Mike Cabral discovered problems over and above the published errors he had been assigned to investigate. For instance, after looking at competing magazines he decided that *Megacrunch* needed improved page design, along with a higher quality stock (the paper the magazine is printed on). He also concluded that a monthly section on business applications would help. But despite their usefulness, none of these findings or ideas was part of Mike's *original* assignment. White expected him to answer these questions, specifically:

- Which errors recur most frequently in our publication?
- Where are these errors coming from?
- What can we do to prevent them?

Mike therefore decides to focus exclusively on the error problem. (He might later discuss those other issues with White—if the opportunity arises. But if *this* report were to include material that exceeds the assignment *and* the reader's expectations, Mike could end up appearing arrogant or presumptuous.)

Mike tries to give White only what she requested. He analyzes the problem and the causes, and then recommends a solution. Mike adapts the general outline on page 529 to shape the three major sections of his report: *introduction, findings and conclusions/recommendations.* For the user's convenience and orientation, he includes the report supplements discussed in Chapter 16: *front matter* (title page, transmittal letter, table of contents, and informative abstract) and *end matter,* as needed (a works cited page [as discussed in Appendix A] and appendixes [not shown here]).

After several revisions, Mike submits copies of the report to his boss and to his instructor.

Virtually any long report has a title page

A forecasting title

Quality-Control Recommendations for *Megacrunch* Magazine

The primary user's name, title, and organization

Prepared for
Marcia S. White
Production Manager
Rangeley Publications

Author's name

by
Michael T. Cabral

Submission date

December 14, 19xx

Master Sheet 145

82 Stephens Road
Boca Grande, FL 08754
December 14, 19xx

Marcia S. White, Production Manager and Vice President
Rangeley Publications, Inc.
167 Dolphin Ave.
Englewood, FL 08567

Dear Ms. White:

Here is my report recommending quality-control measures for *Megacrunch* magazine. The report briefly discusses the history of our quality-control problem, identifies the types of errors we are up against, analyzes possible causes, and recommends three realistic solutions.

My research confirmed exactly what you had feared. The problem is big and deeply rooted: our authors have legitimate complaints; our readers justifiably want information they can put to work; and developers and advertisers have the right to demand fair and complete representation. As a result of client dissatisfaction, competing magazines are gaining readers and authors at our expense.

To have an immediate effect on our quality-control problem, we should act now. Because of our limited budget, I have tried to recommend low-cost, high-return solutions. If you have other solutions in mind, I would be happy to research them for projected effectiveness and feasibility.

Sincerely,

Michael T. Cabral

Michael T. Cabral
Production Assistant
Rangeley Publications

Master Sheet 146

CONTENTS

Table of contents (reports with numerous visuals also have a table of figures)
Front matter (items that precede the report)

Heads and subheads from the report itself

All heads in the table of contents follow the exact phrasing of those in the report text

The various typefaces and indentations reflect the respective rank of various heads in the report

Each head listed in the table of contents is assigned a page number

End matter (items that follow the report)

Master Sheet 147

INFORMATIVE ABSTRACT

The informative abstract summarizes the report's essential message (findings, conclusions, recommendations). This is the one part of a long report read most often

The summary stands alone in meaning—a kind of mini report written for laypersons

Busy audiences need to know quickly what is important. A summary gives them enough information to decide whether they should read the whole report, parts of it, or none of it

An investigation of the quality-control problem at *Megacrunch* magazine identifies not only the types of errors and their causes, but also recommends a plan.

Megacrunch suffers from the following avoidable errors:

- *Grammatical errors* are most frequent: misspellings, fragmented and jumbled sentences, misplaced punctuation, and so on.
- *Informational errors:* incorrect prices, products attributed to wrong companies, mismarked visuals, and so on.
- *Technical errors* are less frequent, but the most dangerous: garbled source code, mismarked diagrams, misused technical terms, and so on.
- *Distortions of the author's original meaning:* introduced by editors who attempt to improve clarity and style.

The above errors seem to have the following causes:

- *Poor initial submissions from contributors* ignore basic rules of grammar, clarity, and organization.
- *Lack of structure in the editing cycle* allows for unrestrained and often excessive editing at all stages.
- *Lack of diversity in the editorial staff* leaves language specialists responsible for catching technical and informational errors.
- *Lack of communication with authors and advertisers* leaves the primary sources out of the production process.

On the basis of my findings, I offer three recommendations for improving quality control during the production process:

- *Expanded author's guide* that includes guidelines for effective use of active voice, visuals, direct address, audience analysis, and so on.
- *Five-stage editing cycle* that specifies everyone's duties at each stage. The cycle would require two additional staff members: a technical editor and a fact checker/typist for editorial changes.
- *More communication with contributors* by exchanging galley proofs and increasing our use of the electronic network.

Master Sheet 148

INTRODUCTION

The section tells what the report is about, why it was written, and how much it covers

An overview of the problem and its effects on the magazine's revenues

The audience is referred to appendixes for details that would interrupt the report flow

Request for action

Purpose and scope of report; overview of research methods and data sources

Because his primary audience knows the background, Mike keeps the introduction brief

The reputation of *Megacrunch* magazine is jeopardized by grammatical, technical, and other errors appearing in each issue.

Megacrunch has begun to lose some long-time readers, advertisers, and authors. Although many readers continue to praise the usefulness of our information, complaints about errors are increasing and subscriptions are falling. Advertisers and authors increasingly point to articles or layouts in which excessive editing has been introduced, and some have taken their business and articles to competing magazines. One disgruntled subscriber sums up our problem by asking that we devote "more effort to publishing a magazine without the kinds of elementary errors that distract readers from the content" (Grendel). This kind of complaint is typical of the sample letters in Appendix A.

Granted, complaints are inevitable—as can be seen in a quick review of "Letters to the Editor" in virtually any publication. But if *Megacrunch* is to withstand the competition and uphold its reputation as the leading resource for Cosmo applications in small business, we must minimize such complaints.

This report identifies the major errors that recur in our magazine, and investigates their causes. My data is compiled from interviews with our editorial staff, a review of complaint letters from authors and readers, and a spot-check for errors in the magazine itself. Books and articles on writing and editing provide theory and technique. The report concludes by recommending a three-part solution to our error problem.

Master Sheet 149

FINDINGS AND CONCLUSIONS

ELEMENTS OF THE PROBLEM

Errors in *Megacrunch* are limited to no single category. For example, some errors are tied to technical slip-ups, while others result from editors changing the author's intended meaning. My spot-check of *Megacrunch* 8.10, our most recent issue, revealed errors of the types listed in Table 1.

TABLE 1 Sample Errors

Spot-check of *Megacrunch* 8.10		
Error type	**As Published**	**Corrected Version**
mechanical	<u>varity</u> of software	<u>variety</u> of software
technical	<u>Dos</u>	<u>DOS</u>
informational	<u>Deluxe Panel</u>	<u>DeluxePanel</u>
grammatical	<u>This</u> will help	<u>Editing</u> will help
grammatical	. . . everyone helps <u>for of</u> a program's release date approaches. . . .	. . . everyone helps <u>as</u> a program's release date approaches. . . .
technical	<u>ram</u>	<u>RAM</u>
informational	<u>cosmo</u>	<u>Cosmo</u>
mechanical	We're back, now we will	We're back; now we will

My random analysis of only six pages identified errors in four categories: grammatical/mechanical, technical, informational, and distortions of intended meaning.

Errors in Grammar and Mechanics

Basic correctness is a "given"—and a problem—for any publication. Sentence fragments, confused punctuation, and poor spelling cause readers to "question the professionalism or diligence of both authors and editor" (Chang 39). Even worse, such errors serve as "evidence of ignorance or sloppiness" (Samson 10). These assertions are borne out by the sampling of complaints in Appendix A.

A recent survey of college and workplace writing (Haswell 168) found that the average writer suffers from the following basic problems:

Margin notes (left column):

This section tells what was found and what it means

The first subsection analyzes the problem; the second will examine causes

Introduction to the problem, and a lead-in to the visual

A visual that illustrates parts of the problem

To further segment the report, each major section (INTRODUCTION, FINDINGS AND CONCLUSIONS, RECOMMENDATION) begins on its own separate page

Discussion of the visual, and overview of the subsection

One part of the problem defined, with examples and effects of the problem Citing authorities clarifies and supports the author's position

3

Author interprets and relates the material to the users, who want to know what this means to them personally

Request for action

- Three-to-four words are misspelled in a memo-length piece.
- One of every ten sentences is a run-on or an "attachable sentence fragment."
- Every fifth possessive is incorrectly formed.

Given these findings, we should not be surprised to receive imperfect manuscripts. However, we must eliminate the slips of the so-called average writer before final copy goes to press.

Another part of the problem defined

Examples

Effects of the problem

Errors in Information Accuracy and Access

Beyond basic errors, we have published some inaccurate information. For instance, we sometimes attribute products to the wrong companies or we list incorrect prices. Inaccuracies of this kind infuriate readers, product developers, and suppliers alike. And a retraction printed in the magazine's subsequent issue has little impact once the damage has been done.

**Other examples
The need for action**

Besides inaccurate information, *Megacrunch* too often presents inaccessible information. Mismarked visuals, misplaced headings, and misnumbered page references make the magazine hard to follow and use selectively.

Another part of the problem defined

Effects of the problem

Examples

Technical Errors

Technical errors seem one of our biggest problems. While some readers might raise a proverbial eyebrow over grammatical errors or skim over informational errors, technical errors are more frustrating and incapacitating. On a page of text, a misplaced comma or a missing bracket can be irritating, but in a program listing, these same errors can render the program useless. Even worse, a misnamed or misnumbered pin or socket in a hardware diagram might cause users to inadvertently destroy their data or damage their hardware.

Some technical slips in *Megacrunch* have veered close to disaster. Consider, for example, the flawed diagram in Figure 1, from our 7.12 issue.

A vivid example, as a visual

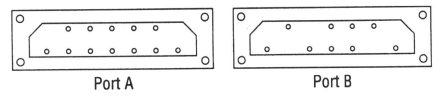

Port A   Port B

To connect the external drive, plug Cable Y into Port A

FIGURE 1. Partial View of the Port Panel on the AXL 100

Master Sheet 151

Our published diagram instructed users to plug a 9-pin external-drive cable into Port A, a 12-pin modem port; the correct connection was to have been made to Port B. Ron Catabia, author of the article and respected tech wizard, explained the flaw in our reproduction of his diagram: "Had any users followed the instructions as printed, the read/write head on their external drive could have suffered permanent damage" (Catabia). Our lengthy correction printed one month later was in no way a sufficient response to an error of this importance. Nor could our belated correction placate an enraged and discredited author.

Such errors do little to encourage reader's perception of *Megacrunch* as the serious user's resource for the latest technical information.

Distortions of the Author's Original Meaning

Authors routinely complain that, in our efforts to increase clarity and readability, we distort their original, intended meaning. After reading the edited version of her article, one author insisted that "too often, edits actually changed what I had said to something I hadn't said—sometimes to the point of altering the facts" (Dimmersdale).

Editorial liberties inevitably alienate authors. Overzealous editors who set out to shorten a sentence or fine-tune a clause—while knowing nothing about the program being discussed—can distort the author's meaning. As a recent study confirms, "When reviewers [editors] criticize in areas outside their expertise, their misguided reviews are seen as an intrusion" (Barker 37).

Following is an excerpt that typifies the distortions in recent issues of *Megacrunch*. Here, a seemingly minor editorial change (from "but" to "even") radically changes the meaning.

As submitted: A user can complete the Filibond program without ever having typed but a single command.

As published: A user can complete the Filibond program without ever having typed even a single command.

As the irate author later pointed out, "My intent was to indicate that a single command must be typed during the program run" (Klause).

This type of wholesale editing (of which more examples are shown in Appendix B) is a disservice to all parties: author, reader, and magazine.

CAUSES OF OUR EDITORIAL INACCURACY

Before devising a plan for dealing with our editing difficulties, we have to answer questions such as these:

Discussion of the visual

Effects of the error

Interpretation—what it all means for the magazine

Another part of the problem defined

Example

Interpretation

Citing an authority to support the interpretation

Another example

Effect of the error

Second subsection

Justification for an analysis of causes

Master Sheet 152

- Where are these errors coming from?
- Can they be prevented?

Interviews with our editing staff along with analysis of our editing practices and review of letters on file uncovered the following causes: (1) poor initial submissions from contributors, (2) lack of structure in our editing cycle, (3) lack of diversity in our editing staff, and (4) lack of communication with authors and advertisers.

Poor Initial Submission from Contributors

Some contributors submit poorly written manuscripts. And so we edit heavily whenever "a submission otherwise deserves flat-out rejection," as one editor argues. Our editors claim that printing poor writing would be more damaging than the occasional editing excesses that now occur. Although editors can improve clarity and readability without in-depth knowledge of the subject, we often misinterpret the author. Clearer writing guidelines for authors would result in manuscripts needing less editing to begin with. Our single-page author's guide is inadequate.

Lack of Structure in Our Editing Cycle

In our current editing cycle, the most thorough editors see an article repeatedly, as often as time allows. Various editors are free to edit heavily at all stages. And these editors are entirely responsible for judgments about grammatical, informational, and technical accuracy.

Although "having your best give their best" throughout the cycle seems a good idea, this approach leads to inconsistent editing and/or overediting. Some editors do a light editing job, choosing to preserve the original writing. Others prefer to "overhaul" the original. With light-versus-heavy editing styles entering the cycle randomly, errors slip by. As one editor noted, "Sometimes an article doesn't get a tough edit until the third or fourth reading. At that point, we have no time to review these last-minute changes" (*Megacrunch* editing staff).

Any article heavily edited and rewritten in the final stages stands a chance of containing typographical and mechanical errors, some questionable sentence structures, inadvertent technical changes, and other problems that result from a "rough-and-tumble edit."

Although some articles are edited inconsistently, others are overedited. Our editors tend to be vigilant in pursuit of clarity, conciseness, and tone. Unfortunately, they seem less vigilant about technical accuracy.

Margin annotations (left column):

- Scope of this subsection, so that users know what to expect
- First cause defined
- Findings
- Conclusion
- Second cause defined
- Findings
- Interpretation— what it means
- Findings
- Interpretation

Master Sheet 153

Conclusion

Instead of full-scale editing at all stages, we need a cycle that makes a manuscript progress from inadequate (or adequate) to excellent, through different levels of editorial attention. For example, a first edit should be thorough, but a final proofreading should be merely a fine-combing for typographical and mechanical errors.

Third cause defined and interpreted

Lack of Diversity in Our Editing Staff

The variety of errors suggests that our present staff alone cannot spot all problems. Strong writing backgrounds have not prepared our editors to recognize a jumbled line of programming code or a misquoted price. To snag all errors, we must hire technical specialists. We need both a technical editor and a fact checker, to pick up where current editors leave off.

Conclusion

Fourth cause defined and interpreted

Lack of Communication with Authors and Advertisers

Some of our editing troubles emerge from a gap between the meaning intended by contributors and the interpretation by editors. In the present system, contributors submit manuscripts without seeing any editorial changes until the published version appears. Along with an expanded author's guide, regular communication throughout the editing process (and perhaps the writing process as well) would involve contributors in developing the published piece, and thus make authors more responsible for their work.

Conclusion

Master Sheet 154

RECOMMENDATIONS

To eliminate published inaccuracies, I recommend: (1) an expanded author's guide, (2) a five-stage editing cycle, (3) a checklist for each stage, and (4) improved communication with contributors.

EXPANDED AUTHOR'S GUIDE

The obvious way to limit editing changes would be to accept only near-perfect submissions. But as a technical resource we cannot afford to reject poorly written articles that are nonetheless technically valuable.

To reduce editing required on submissions, I recommend we expand our author's guide to include topics like these: audience analysis, use of direct address and active voice, principles of outlining and formatting, and use of visuals.

FIVE-STAGE EDITING CYCLE

In place of haphazard editing, I propose a progressive, five-stage cycle: Stages one through three would refine grammar, clarity, and readability. Two additional staff members, a fact checker and a technical editor, would check facts and technical accuracy in the final two stages. Figure 2 outlines responsibilities at each stage.

A CHECKLIST FOR EACH STAGE

To ensure a consistent focus throughout the editing cycle, our staff should collaborate immediately to develop a detailed checklist for each stage (Hansen 15). As a guide for editors as well as for contributors, these checklists would enhance communication among all parties.

IMPROVED COMMUNICATION WITH CONTRIBUTORS

The following measures would reduce errors caused by misunderstandings between contributors and editors.

Author-Client Verification of Galley Proofs

Two weeks before our deadline, we could send authors pre-publication galley proofs [which show the text as it will appear in published form]. Authors could check for technical errors or changes in meaning, and return proofs within five days.

Expanded Use of Our Electronic-Mail Network

At any time during production, authors and editors could communicate through CompuServe by leaving questions and messages in one another's electronic mailboxes. (Virtually all our regular authors subscribe to CompuServe.) In addition, all parties could log onto CompuServe at one or more scheduled times daily to discuss the manuscript. The E-mail alternative is cheaper than the telephone, eliminates "telephone tag," and could serve as a "hot line" for authors while they prepare a manuscript for submission.

Master Sheet 155

The visual illustrates and summarizes the process being recommended

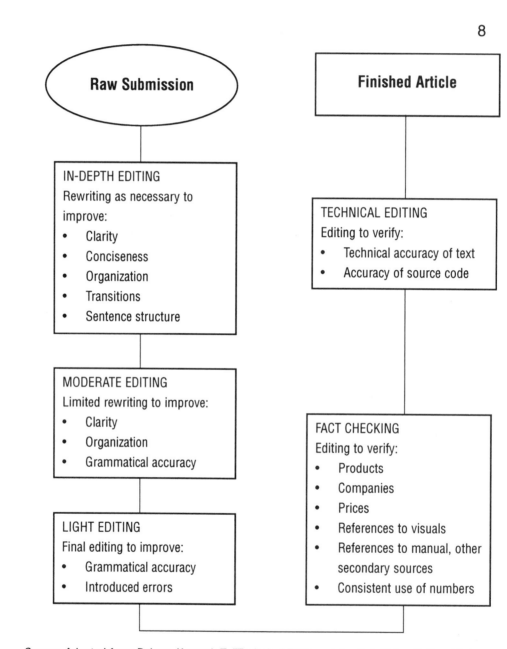

Data sources appear directly below the visual

Source. Adapted from Rainey, Kenneth T. "Technical Editing at the Oak Ridge National Laboratory. " *Journal of Technical Writing and Communication* 18 (1988): 175–81 and from Unikel, George. "The Two-Level Concept of Editing." *The Technical Writing Teacher* XV (1988): 49–54.

FIGURE 2 The Five-stage Editing Cycle

Master Sheet 156

Possible Use of Groupware

Third part of final recommendation

For unified and efficient collaboration, we should also explore the feasibility of groupware (software, such as *ForComment* or *MarkUp,* for collaborative editing online). Even though research indicates that editing is best done on hard copy pages rather than on a computer screen (Kaufer and Neuwirth 113–15), groupware would provide a record of all edits at all stages of the editing cycle. Such a record would be invaluable in helping us troubleshoot and refine our editing process.

Master Sheet 157

WORKS CITED

The list of works cited clearly identifies each source cited in the report

Barker, T. "Feedback in High-Tech Writing." <u>Journal of Technical Writing and Communication</u> 18.1 (1988): 35–51.

Catabia, R. Notes from author Catabia's phone conversation with the Managing Editor. 10 Mar. 1999.

Chang, F. "Revise: A Computer-Based Writing Assistant." <u>Journal of Technical Writing and Communication</u> 17.1 (1987): 25–41.

Dimmersdale, O. Author's letter to the Managing Editor. 6 July 1999.

Grendel, M.L. Subscriber's letter to the Managing Editor. 14 May 1999.

Hansen, James B. "Editing Your Own Writing." <u>Intercom</u> Feb. 1997: 14–16.

Haswell, R. "Toward Competent Writing in the Workplace." <u>Journal of Technical Writing and Communication</u> 18.2 (1988): 161–72.

Kaufer, David S., and Chris Neuwirth. "Supporting Online Team Editing: Using Technology to Shape Performance and to Monitor Individual and Group Action." <u>Computers and Composition</u> 12 (1995): 113–24.

Klause, M. Author's letter to the Managing Editor. 11 Nov. 1999.

<u>Megacrunch</u> editing staff. Interviews. 12–14 April 1999.

Specific names omitted from this citation, to protect in-house sources and to allow employees to speak candidly, without fear of reprisal

Samson, Donald C. <u>Editing Technical Writing</u>. New York: Oxford, 1993.

288